BIOSEPARATIONS

Principles and Techniques

B. SIVASANKAR
Professor
Department of Chemistry
Anna University
Chennai

PHI Learning Private Limited
Delhi-110092
2019

₹ 325.00

BIOSEPARATIONS: PRINCIPLES AND TECHNIQUES
B. Sivasankar

ISBN-978-81-203-2649-1

Twelfth Printing ... ... June, 2019

Published by Asoke K. Ghosh, PHI Learning Private Limited, Rimjhim House, 111, Patparganj Industrial Estate, Delhi-110092 and Printed by Mudrak, D-61, Sector 63, Noida, U.P.-201301.

Contents

Preface *vii*

1. **AN OVERVIEW OF BIOSEPARATIONS** **1–12**
   1.1 Bioprocesses *1*
   1.2 Range and Characteristics of Bioproducts *2*
   1.3 Need for Downstream Processing *3*
   1.4 Characteristics of Fermentation Broths *4*
   1.5 An Overview of Bioseparations *8*
   1.6 A Few Case Studies *10*
   Questions 12

2. **CELL DISRUPTION** **13–25**
   2.1 Intracellular Products *13*
   2.2 Cell Wall *13*
   2.3 Cell Disruption *15*
   2.4 Proteins of Inclusion Bodies *24*
   Questions 25

3. **FILTRATION** **26–41**
   3.1 Solid–Liquid Separation *26*
   3.2 Theory of Batch Filtration *27*
   3.3 Pretreatment of Fermentation Broths *30*
   3.4 Filter Media *32*
   3.5 Equipment *32*
   3.6 Washing of Filter Cakes *33*
   3.7 Continuous Filtration *34*
   Exercises 37
   Questions 41

4. **CENTRIFUGATION** **42–53**
   4.1 Centrifugal Force *42*
   4.2 Centrifugal Sedimentation *43*
   4.3 Centrifugal Decantation *45*

4.4 Centrifuges *45*
4.5 Selection of Centrifuge Type *49*
4.6 Scale-up *50*
4.7 Centrifugal Filtration *50*
Exercises 52
Questions 53

5. ADSORPTION **54–66**
5.1 Adsorption Process *54*
5.2 Adsorption Isotherms *54*
5.3 Adsorption Techniques *56*
Exercises 64
Questions 66

6. EXTRACTION **67–99**
6.1 Liquid–Liquid Extraction *67*
6.2 Solvent Extraction Principles *68*
6.3 Extraction Process *71*
6.4 Equipment for Extraction *74*
6.5 Operating Modes of Extraction *76*
6.6 Aqueous Two-phase Extraction *81*
6.7 Theoretical Principles of Aqueous Two-phase Extractions *85*
6.8 Aqueous Two-phase Extraction Process *86*
6.9 Equipment for Aqueous Two-phase Extraction *87*
6.10 Applications of Aqueous Two-phase Extraction *88*
6.11 Reversed Micellar Extraction *90*
6.12 Supercritical Fluid Extraction *91*
Exercises 96
Questions 98

7. MEMBRANE SEPARATION PROCESSES **100–118**
7.1 Merits of the Process *100*
7.2 Classification of Membrane Separation Processes *101*
7.3 Theoretical Models for Membrane Processes *102*
7.4 Retention Coefficient or Rejection Coefficient *103*
7.5 Factors Affecting the Separation Processes *103*
7.6 Operational Requirement of Membranes *106*
7.7 Structure of Membranes *106*
7.8 Preparation of Membranes *107*
7.9 Equipment *108*
7.10 Microfiltration or Cross-flow Filtration *110*
7.11 Ultrafiltration *112*
7.12 Reverse Osmosis (RO) (Hyperfiltration) *114*
7.13 Dialysis *116*
7.14 Electrodialysis *117*

7.15 Pervaporation *117*
7.16 Applications of Membrane Separation Processes *118*
Questions 118

8. PRECIPITATION 119–142
8.1 Precipitation of Proteins *119*
8.2 Protein Precipitation Methods *120*
8.3 Selective Denaturation of Unwanted Proteins *133*
8.4 Large Scale Precipitation (Process Scale-up) *137*
Exercises 141
Questions 142

9. CHROMATOGRAPHY: PRINCIPLES AND PRACTICE 143–174
9.1 Chromatogrphy—A Separation Technique *143*
9.2 Classification of Chromatographic Techniques *143*
9.3 General Description of Column Chromatography *145*
9.4 Chromatographic Terms and Parameters *148*
9.5 Practice of Chromatography *161*
9.6 HPLC *163*
9.7 Scale-up of Chromatography *166*
9.8 Planar Chromatographic Techniques *167*
9.9 Process Considerations in Preparative Liquid Chromatography *169*
Exercises 171
Questions 174

10. GEL FILTRATION 175–187
10.1 Size Exclusion Chromatography *175*
10.2 Basic Principles *175*
10.3 Relationship between Molecular Size of Solute and Retention in Gel Filtration *179*
10.4 Materials for Gel Filtration *180*
10.5 Equipment *181*
10.6 the Procedure *181*
10.7 Applications *182*
Exercises 186
Questions 187

11. ION EXCHANGE CHROMATOGRAPHY AND CHROMATOFOCUSING 188–199
11.1 Principle *188*
11.2 Ion Exchangers *189*
11.3 Capacity of Ion Exchangers *190*
11.4 Operating Modes *192*
11.5 Practice of Ion Exchange Chromatography (IEC) *193*
11.6 Chromatofocusing *196*
Questions 199

12. REVERSED PHASE AND HYDROPHOBIC INTERACTION CHROMATOGRAPHY **200–213**

12.1 Hydrophobic Interaction *200*
12.2 Reversed Phase Chromatography and Hydrophobic Interaction Chromatography *201*
12.3 Basic Theory of Retention in RPC and HIC *202*
12.4 Reversed Phase Chromatography *203*
12.5 Hydrophobic Interaction Chromatography *209*
Questions *213*

13. AFFINITY CHROMATOGRAPHY **214–239**

13.1 General Features *214*
13.2 Specific and Non-specific Interactions *214*
13.3 Bioaffinity Chromatography *215*
13.4 Immobilization of Ligands *224*
13.5 Pseudoaffinity Chromatography *227*
13.6 Immobilized Metal Ion Affinity Chromatography (IMAC) *230*
13.7 Covalent Chromatography *232*
Questions *239*

14. ELECTROKINETIC METHODS OF SEPARATION **240–254**

14.1 the Various Methods *240*
14.2 Electrophoresis *240*
14.3 Capillary Electrophoresis *248*
14.4 Isoelectric Focussing *250*
14.5 Isotachophoresis *252*
Questions *254*

15. FINISHING OPERATIONS AND FORMULATION **255–265**

15.1 Finishing Operations *255*
15.2 Crystallization *255*
15.3 Drying *259*
15.4 Formulation *264*
Questions *265*

BIBLIOGRAPHY ***267***

INDEX ***269–271***

Preface

In the last few years biotechnology is gaining importance in India and has been recognized as a thrust area in teaching, research and industrial practice. Biotechnology and biochemical engineering programmes are currently offered at the graduate (B.Sc. and B.Tech.) as well as at the postgraduate (M.Sc. and M.Tech.) level by various universities in India.

Bioseparations also called *downstream processing* in biotechnology is a core course both in science and technology programmes at the graduate as well as at the postgraduate levels. Downstream processing is an essential and important activity in biotechnological industries by means of which a wide variety of products such as pharmaceuticals, drug intermediates, food chemicals, beverages, organic fine chemicals and solvents, industrial enzymes, dairy products etc. are manufactured. A varierty of techniques are adopted to isolate and recover the desired industrial product from a complex mixture of starting materials, reaction products and by-products, to concentrate the desired product, to purify it and finally to formulate the product for long term storage, transporting and marketing before it reaches the end use customer. The techniques adopted include conventional chemical engineering unit operations as well as sophisticated high resolution techniques. Suffice it to say that biotechnology without downstream processing would be incomplete.

This book aims at providing comprehensive information on the topics of importance to students of biotechnology and biochemical engineering. The book deals with the theoretical principles involved in the techniques adopted both at the research laboratory level and manufacturing scale. The various topics have been covered in 15 chapters based on the different stages of downstream processing practice, which have been outlined in Chapter 1. The release of intracellular products produced is an important phase of downstream processing involving different methods of cell wall disruption as described in Chapter 2. Solid–liquid separation by filtration and centrifugation have been discussed in Chapters 3 and 4 respectively. Major product isolation and concentration methodologies of adsorption, extraction and membrane separation as practiced in industry have been highlighted in Chapters 5–7. The importance of precipitation as a technique of isolation and purification of proteins has been related in Chapter 8. Chapter 9 has been written with regard to the fact that students of biology, chemistry, pharmaceutical science/technology and biochemical engineering require the theoretical principles of chromatography as a high resolution technique which can be adopted for laboratory as well as industrial scale operations. Related chromatographic techniques such as gel filtration, reversed phase and hydrophobic interaction chromatography, affinity and pseudoaffinity chromatography, ion-exchange chromatography and other methods have been discussed in Chapters 10–13. Chapter 14 is devoted to electrokinetic methods of separation while Chapter 15 describes various finishing operations.

The material for the book has been prepared with due consideration to the diverse requirements of students who wish to specialize in different areas of biotechnology. The book fulfils the syllabi requirements of most of the university programmes in biotechnology. The book will be useful as a textbook for the core course on downstream processing/bioseparation technology taken by undergraduate, postgraduate and research students in biotechnology and pharmaceutical science/technology as well as to students of biochemistry and chemical engineering.

It is a pleasant privilege to acknowledge the contributions of my teachers, well wishers and colleagues in bringing out this book. I take this opportunity to express my heart felt thanks to (Late) Prof. B. Jagannadhaswamy and Prof. C.M. Lakshmanan, former Directors of A.C. College of Technology, Anna University, Chennai, for their encouragement and guidance, Prof. Kunthala Jayaraman of Center for Biotechnology, Anna University, Chennai and Prof. M.A. Vijayalakshmi of University of Technology, Compiegne, France for initiating me into this fascinating subject. I express my special thanks to Prof. S.V. Raman, Visiting Faculty, Anna University for his enlightening discussions, clarifications and editing of the manuscript of chapters on filtration, centrifugation and adsorption. My friends and colleagues Prof. K. Rengaraj and Prof. V. Sadasivam of Chemistry Department, Anna University have always had the kindness and time to encourage me. Authorities of Anna University and my colleagues in the Chemistry Department deserve special mention for the support they always provide for academic pursuits. I am at a loss for words in expressing my appreciation and gratitude to my family members, especially my mother and wife for their patience and understanding and my children for their support. Finally, I convey my sincere thanks to the publishers, Prentice-Hall of India, especially to Mr. Sakthivel of their marketing division for his perseverance and their editorial and production division for their wholehearted efforts in bringing out this book.

B. SIVASANKAR

CHAPTER

1

An Overview of Bioseparations

1.1 BIOPROCESSES

Bioprocesses make use of living cells and microorganisms as well as enzymes for the production of a variety of bulk organic chemicals, food products, pharmaceuticals and fuels.

The salient features of biotechnological processes include:

- Use of ambient conditions of temperature and pressure.
- High product specificity, particularly in the production of stereo specific and enantiospecific organic chemicals, drugs, fungicides and pesticides.
- Relatively clean technology with respect to environmental management.
- Bioprocess is possibly the only viable route for the production of high-value low-volume specialty products such as therapeutic and diagnostic enzymes and monoclonal antibodies.
- Other advantageous features over conventional manufacturing processes include the use of natural renewable feedstock such as carbohydrates as a source of carbon for the growth and sustenance of metabolic activity of the microorganisms during fermentation to produce useful products. In contrast, chemical industries depend on the non-renewable and depleting fossil fuels (mainly petroleum) for their feedstock.

However, bioprocesses are also associated with a few major disadvantages such as the formation of a multitude of products during fermentation, some of which may affect the stability or bioactivity of the desired products and the very low concentrations of desired products which necessitates handling of large volumes of fermentation broths. The disadvantages often call for the use of complex downstream processing steps.

1.2 RANGE AND CHARACTERISTICS OF BIOPRODUCTS

Bioproducts include a wide range of chemicals, which may be broadly classified into three major categories on the basis of market volume, market price and requirement of purity, as shown in Figure 1.1. These include (a) very high value, low volume products such as therapeutic proteins and enzymes, factor VIII, interferon, urokinase etc. of very high purity the volume produced being in the range of grams to kilogram; (b) high value, low volume, high purity products such as diagnostic enzymes (e.g. luciferase and glycerophosphate dehydrogenase), human growth hormone, tissue plasminogen activator, monoclonal antibodies and insulin produced in tens or hundreds of kilograms; and (c) bulk industrial products of relatively low purity such as organic acids, amino acids, ethanol, antibiotics, proteases and amylases produced in hundreds of kilograms to tons in quantity.

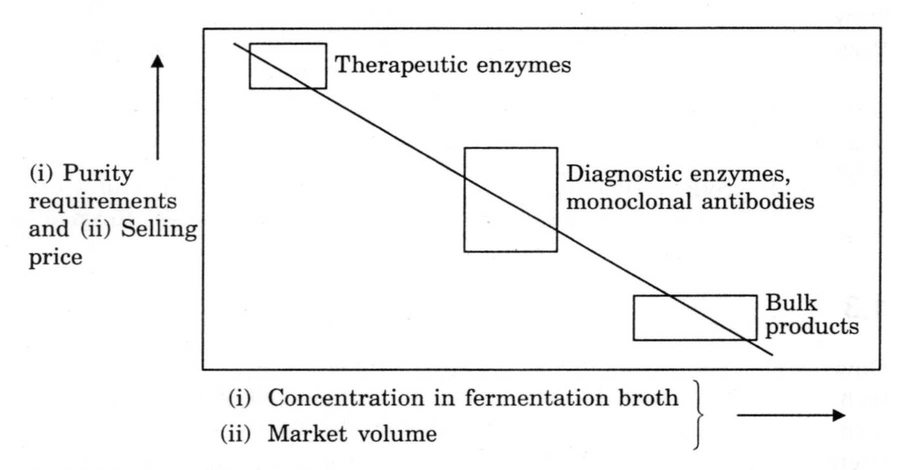

Figure 1.1 Correlation between market volume or concentration of the desired product in the broth and selling price or purity requirements.

Bioproducts differ greatly in their nature and hence different separation principles and mechanisms depending on molecular mass, charge distribution, hydrophobicity, structure and immunogenic structure and specific affinity towards other biomolecules become necessary for their isolation and purification. The important characteristics of bioproducts include molecular weights and fragility. The molecular masses vary from about 50 to 2,000,000. Most of the bioproducts are fragile and are destabilized by factors such as variations in pH, temperature, ionic strength and the nature of solvent. In addition, shear forces during stirring and the presence of denaturing agents such as surfactants, metal ions and other chemicals affect the stability and bioactivity of the products. Thus the operating conditions of bioprocesses depend on the characteristics of bioproducts.

Many of the products of bioprocess industries are produced by batch methods in relatively small scale level. Most of the high value low volume products are manufactured under Good Manufacturing Practice (GMP) as per the requirements of national regulatory authorities under sub-optimal conditions, not governed by economic considerations, so as to introduce their product first in the market.

The characteristics of bioprocess products are summarized in Table 1.1.

TABLE 1.1 Characteristics of Bioprocess Products

Characteristic feature	*Categories*		
	I	*II*	*III*
Market volume	0.1–100 kg/year	10^3–10^5 kg/year	10^6–10^9 kg/year
Types of organisms	rDNA	rDNA	Natural producers
Purity requirement	Very high	High	Low
Manufacturing practice	GMP (under sub-optimal conditions)	GMP (under sub-optimal conditions)	Validated and proven technology
Cost consideration	Not important	Less important	Important
Major bioseparation techniques	Affinity chromatography, preparative electrophoresis	Membrane separation, adsorption chromatography	Precipitation, filtration, extraction, adsorption

1.3 NEED FOR DOWNSTREAM PROCESSING

Downstream processing (bioseparations) is an essential part of bioprocess technology in that the desired product needs to be isolated, purified and formulated for different end uses. It is a challenging area for biologists, chemists and chemical engineers as the problems involved in the separation of biological products are numerous.

The products are manufactured using a variety of equipment. Besides fermenters or bioreactors, other special reactors such as airlift, membrane and immobilized cell reactors are also used. All the reactors operate under sterile conditions. A variety of microorganisms including genetically engineered species are used for the production of desired products. The products formed may be secreted into the broth or may be retained within the cell introducing additional complexity in the recovery of the product. The products formed are usually in low concentrations necessitating the handling of large volumes and in some cases, the broths are viscous creating additional problems of fouling of equipment.

The downstream processing steps are based on steady as well as unsteady state techniques. The principles of recovery of the product depends on such well studied physical properties such as density, distribution coefficient as well as less known biospecific interactions involving molecular weight, affinity towards specific molecules, charge distribution, hydrophobicity or structure. Some of these steps are exclusive for the individual product.

In the so-called *classical biotechnology*, wherein the products are produced naturally by the living cells as in the case of penicillin by molds, ethanol by yeast and organic acids and enzymes by bacteria, conventional chemical engineering unit operations are useful in the isolation and purification of such products. The unit operations include filtration, centrifugation, sedimentation, adsorption and liquid–liquid extraction. Most of these unit operations are well described mathematically and scale-up parameters are relatively easy to determine through pilot plant studies.

In *modern biotechnology* the microorganisms are induced under specific conditions, some times through recombinant DNA technology, to produce products, which are not natural metabolic products. The recovery of such products requires a number of steps involving high-resolution techniques such as affinity methods of adsorption, extraction, precipitation or chromatography.

The choice of the separation methodology depends to a large extent on the nature of the product, its quantity and the extent of purity required. Before venturing into the selection of appropriate downstream processing steps it is necessary to understand certain basic features of fermentation broths.

1.4 CHARACTERISTICS OF FERMENTATION BROTHS

The characteristics of fermentation broths that influence the downstream processing of biomolecules include (i) the type of microorganisms and their morphological features (size and shape), (ii) concentration of cells, products and byproducts and (iii) physical and rheological characteristics.

1.4.1 Morphology of Cells

A large variety of microbial, plant and animal cells with cells ranging in size from about 1 μm for bacteria to about 4,000 μm for cellular agglomerates are encountered in fermentation broths as shown in Figure 1.2. In addition, non-consumed media components, cell organalle, low molecular weight products such as inorganic ions, sugars, organic acids and antibiotics as well as high molecular weight products such as polysaccharides, DNA and RNA are also found in the broths.

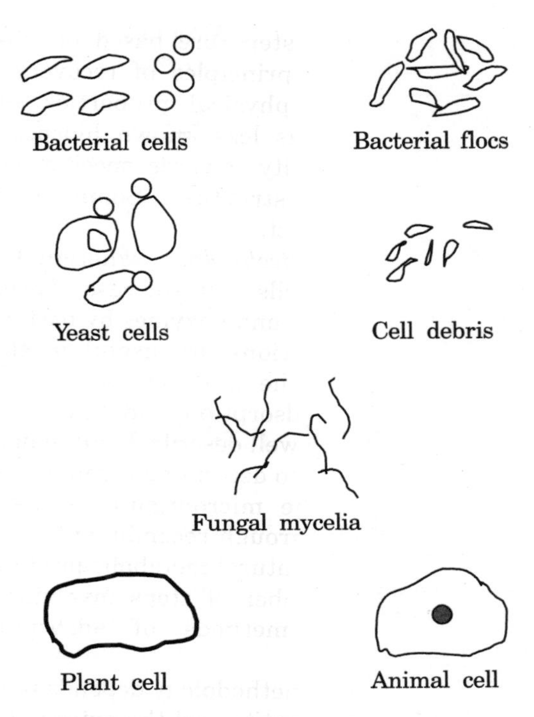

Figure 1.2 Morphology of cells.

The cells and the cell agglomerates exhibit a variety of shapes. Bacterial and yeast cells occur mostly as homogeneously suspended particles in the fermentation broth. Certain strains of bacteria form a slimy mass, which is difficult to separate from the liquid. The viscosity of the liquid also increases due to the presence of a slimy mass. In addition, the slimy mass may clog and foul the equipment. Fungi form a network of hyphi thereby increasing the viscosity of the broth. Under certain conditions fungi form agglomerates called pellets of relatively large size of about 100–4,000 μm, which are easy to separate. In general, the capacity of a separation process decreases with decreasing cell size and consequently the cost involved increases.

1.4.2 Concentration of Cells, Products and Byproducts

The concentrations of the biomass and that of the products in the fermentation broth are important in deciding on the choice of the separation process. The concentrations of the biomass as well as that of the products vary widely as shown in Table 1.2.

TABLE 1.2 Concentrations of Biomass and Products in Fermentation Broths

Biomass	*Concentration (% dry weight)*	*Size* (μm)	*Bioproduct*	*Size* (nm)	*Concentration* (% W/W)
Cell debris	3–6	0.5–2	Alcohols	0.3–0.5	8–12
Bacterial cells	3–6	0.5–3	Organic acids	0.2–0.6	5–10
Yeast	3–6	5–10	Inorganic ions	0.1–0.3	0.1–0.3
Fungi	2–4	10–50	Sugars	0.5–1.0	0.1–0.2
Animal cells	0.02–0.05	10–80	Antibiotics	0.5–1	3–5
Plant cells	0.1–0.5	10–200	Proteins	1–10	0.05–1

1.4.3 Physical and Rheological Characteristics

The density of a dry biomass would be about 1400 kg/m^3. However, the density of the fermentation broth is lower, around 1100 kg/m^3 as the cells have a high water content of about 70–80%. Pellets and flocs have a much lower density—almost comparable to the density of the clarified medium which is, about 1030 kg/m^3 due to water entrapped between the cells—making separation of biomass on density alone difficult.

The rheological property of fermentation broth is of importance for downstream processing in the case of centrifugation and membrane separation. The clarified broth after the removal of biomass is almost water-like in its flow characteristics for most fermentations, except in the case of fermentation of polysaccharides such as xanthan gum. Hence simple Newtonian and non-Newtonian models can describe the situation in most of the fermentation broths. The viscosity (resistance of a liquid to flow) of a broth is strongly influenced by the cell concentration as well as cell shape and to a minor extent by the changes in the concentration of nutrients and excreted metabolites.

The viscosity and flow characteristics of a simple liquid such as water and oil may be conveniently described on the basis of Newtonian model in terms of an applied shear stress (force) as given by Eq. (1.1).

$$\tau = \eta\gamma \tag{1.1}$$

where τ is the shear stress (N m^{-2} or Pa), η is dynamic viscosity (N s m^{-2} or Pa s) and γ is the shear rate (s^{-1}, defined as the relative velocity of adjacent parallel layers in a laminar flow of a liquid under shear force). The shear stress is a linear function of the shear rate in Newtonian fluids and the viscosity η is a proportionality constant independent of shear rate.

The viscosity of a dilute aqueous suspension of cells (fermentation broth) depends on the shape and concentration of cells and is best described by Einstein's equation, applicable for a dilute suspension of particles which show no interaction among themselves, as given by Eq. (1.2).

$$\eta_r = \frac{\eta_s}{\eta_l} = 1 + k\phi \tag{1.2}$$

where η_r, η_s and η_l are the relative viscosity, viscosity of the suspension and viscosity of the liquid respectively, k is a constant depending on the shape of the particles ($k = 2.5$ for spherical particles and varies between 20 and 1200 for rod shaped particles) and ϕ is the volume fraction of the solid particles. Equation (1.2) is valid for dilute suspensions only with $\phi < 0.1$. For more concentrated suspensions Eilers' Eq. (1.3) is more useful.

$$\eta_r = \left\{1 + \frac{0.5\,k\phi}{(1 - \phi/\phi_{max})}\right\}^2 \tag{1.3}$$

where ϕ_{max} is the maximum packing density (about 0.6–0.7 for the type of cells under investigation). Thus a ϕ_{max} of 0.6 means that 60% of the total volume consists of cells when the cells are packed together as closely as can be achieved. The variation of relative viscosity as a function of volume fraction of cells for Eqs. (1.2) and (1.3) are shown in Figure 1.3.

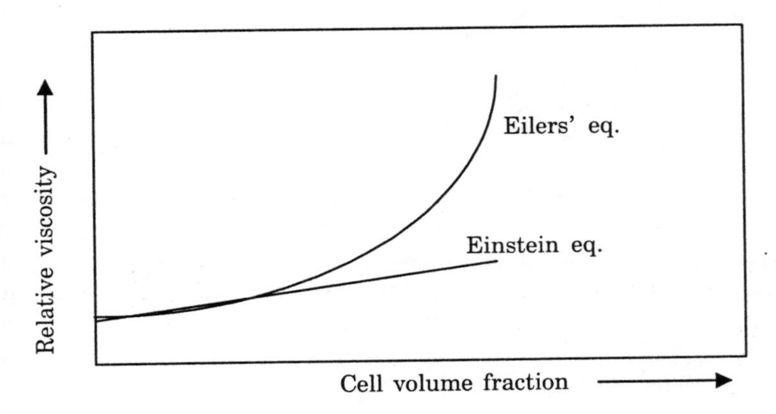

Figure 1.3 Variation of relative viscosity with volume fraction.
Source: P.A. Belter, E.L. Cussler and Wei-Shou Hu, *Bioseparations—Downstream Processing for Biotechnology*, John Wiley & Sons, New York, 1988.

In the case of non-Newtonian fluids, the shear stress is a non-linear function of the shear rate and the power law model may be applied to obtain:

$$\tau = K(\gamma)^n \tag{1.4}$$

where K is consistency index and n is the power law index. Both the parameters can be measured with a rheometer.

The viscosity of mold suspensions in antibiotic or yoghurt fermentation may be characterized by the Bingham model described by:

$$\tau = \tau_0 + \eta\gamma \tag{1.5}$$

where τ_0 is the yield stress which is the minimum stress required to cause liquid flow. The Hershel–Buckley model is a modification relating the τ, τ_0, K and γ as given by Eq. (1.6).

$$\tau = \tau_0 - K(\gamma)^n \tag{1.6}$$

The variation of shear stress as a function of shear rate for both Newtonian and non-Newtonian fluids is shown in Figure 1.4. Curve 2 in the figure shows a dilatant behaviour, i.e., greater viscosity for greater movement which means that viscosity increases with increasing shear stress. Dilatancy is shown by yeast suspensions at higher concentrations (15% dry solids), another example being quick sand. In this case the index n in Eq. (1.4) is larger than 1.

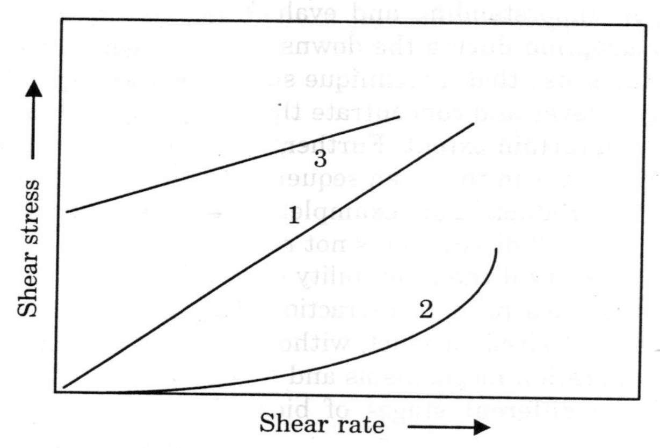

Figure 1.4 Variation of shear stress as a function of shear rate for Newtonian and non-Newtonian fluids. 1. Newtonian, 2. Dilatant, 3. Bingham.

Source: J. Krijgsman, *Product Recovery in Bioprocess Technology*, BIOTOL series, Butterworth-Heinemann Ltd., Oxford, 1992.

The above models can predict the rheological behaviour of fermentation broths, but the actual behaviour should be determined by experiment.

1.5 AN OVERVIEW OF BIOSEPARATIONS

Five stages may be distinguished in the downstream processing as applicable to bioprocess industries. The stages are: (1) product release and pretreatment, (2) removal of insolubles or particulates that is, solid-liquid separation, (3) product concentration and recovery, (4) purification and (5) finishing operations and formulation.

The first stage of product release and pretreatment involves cell disruption to release intracellular products. Pretreatment of fermentation broth by sterilization or pasteurization to stabilize the product and treatment with chemicals to enhance the solid–liquid separation may also be carried out. Filtration or centrifugation in the second stage of downstream processing achieves the removal or separation of particulates. In the third stage, the desired product is concentrated and recovered by using techniques such as adsorption, extraction, precipitation or membrane separation. The purification stage involves the use of high-resolution techniques, mainly based on chromatography and affinity separations. The final stage aims at formulating or finishing the product according to market requirements or for long term storage. The techniques include drying, prilling, tabletting, granulation, crystallization and extrusion.

The above compartmentalization of the stages is purely for academic purposes so as to categorize the various factors and unit operations to be considered for understanding and evaluating the variety of techniques available for adoption during the downstream processing. The stages are not rigid in the sense that a technique such as precipitation or extraction may not only recover and concentrate the desired product but also purify the product to a certain extent. Further, it may not be necessary to carry out all the five stages in the given sequence in the downstream processing of a given bioproduct. For example, if the desired product is an extracellular one, cell disruption is not required and further pretreatment may not be necessary if the filterability of the fermentation broth is good. In certain cases, adsorption or extraction may be adopted directly for the recovery of the desired product without going through the first two stages. The separation mechanisms and the techniques or unit operations available at the different stages of bioseparations are summarized in Table 1.3.

TABLE 1.3 Unit Operations for Different Separation Factors

Separation factor	*Unit operations*
Particle size	Filtration, screening, microfiltration
Molecular size and weight	Ultrafiltration, gel filtration,
Density difference	Ultracentrifugation, centrifugation, cyclone separation, sedimentation
Temperature	Precipitation by thermal denaturation
Diffusivity	Reverse osmosis, dialysis
Solubility	Solvent extraction, precipitation
Ionic charge	Ion-exchange chromatography, electrophoresis
Hydrophobicity	Precipitation, hydrophobic interaction chromatography, reversed phase chromatography
Electrophoretic mobility	Electrophoresis
Isoelectric point	Chromatofocusing, isoelectric focusing
Free thiol groups	Covalent chromatography
Biospecific and biomimetic interactions	Affinity and pseudo-affinity chromatographic techniques

1.6 A FEW CASE STUDIES

The separation steps involved for each product is distinct and unique. In general, as the number of steps increases, the overall yield decreases. For example, if the separation process involves 5 steps each with an yield of 75% the overall yield of the desired product would be 24% only. (Step 1 = 100 × 0.75 = 75%; step 2 = 75% of 75 obtained in step 1, i.e., 56.25%; step 3 = 75% of 56.25% = 42.19% so on). If an overall yield of 50% is required for the same 5-step separation process, the individual step yield should be 87% because the step yield reduction would be greater than 50/5 = 10%, if the step yield is less than 90%. It is usual to compromise yield for purity depending on the nature of the product, market requirement and economic considerations.

The complexity of the separation processes and the uniqueness of the steps involved may be understood with respect to a few typical bioprocesses as given in the flow charts in Figures 1.5 to 1.8.

Fermentation broth

↓ (1) Removal of solids (yeast cells) by screening

Filtrate (in beer still)

↓ (2) Product isolation in distillation column

First distillate

↓ (3) Product purification and polishing in rectifying columns by fractional distillation

Ethanol

Figure 1.5 Broad outline of downstream processing steps in ethanol fermentation.

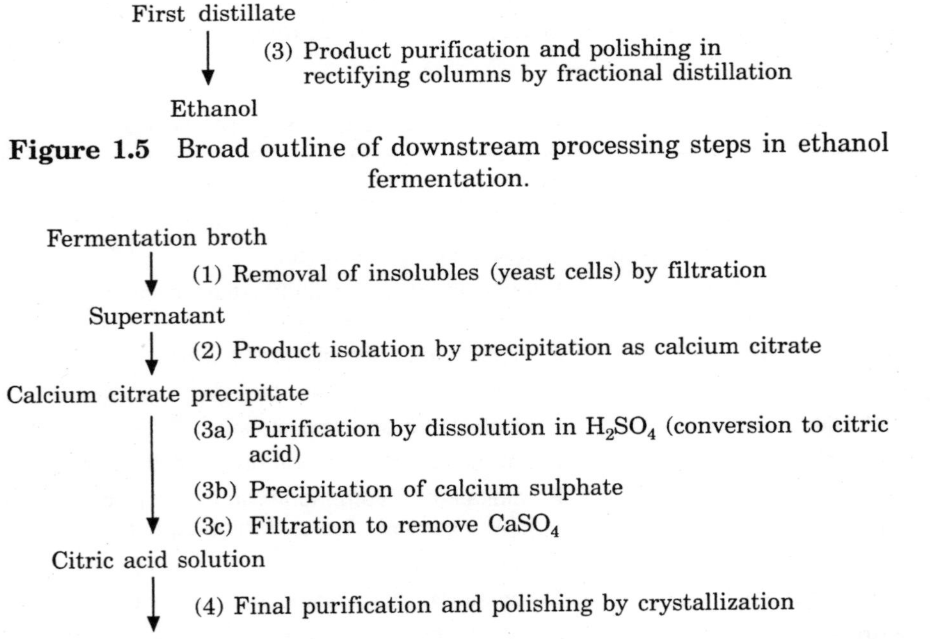

Figure 1.6 Major downstream processing steps in citric acid manufacture.

Fermentation broth

↓ (1) Biomass recovery by centrifugation

Cells (containing intracellular enzyme)

↓ (2a) Product release by homogenization
(2b) Filtration to remove cell debris

Crude enzyme extract

↓ (3) Product isolation by salt mediated precipitation

Crude enzyme precipitate

↓ (4a) Product purification by ultrafiltration and
(4b) Chromatography

Pure enzyme solution

↓ (5a) Final polishing by solvent mediated precipitation
(5b) Solvent removal by dialysis and
(5c) Freeze drying

Finished product

Figure 1.7 Downstream processing steps in the production of an intracellular enzyme.

Fermentation broth

↓ (1) Removal of solids (cells) by filtration

Filtrate (containing antibiotic)

↓ (2) Product isolation by solvent extraction

First extract

↓ (3a) Product purification by stripping
(3b) Second extraction
(3c) Stripping

Pure aqueous extract

↓ (4) Polishing by crystallization and drying

Pure product

Figure 1.8 Downstream processing steps in the production of an antibiotic.

Questions

1. What are the salient features, advantages and disadvantages of bioprocesses compared to conventional chemical processes?
2. What are the characteristic features of bioprocesses?
3. How do biomolecules differ from other chemicals?
4. Give a detailed account of the characteristics of fermentation broths.
5. Discuss the different models describing the flow characteristics of fluids.
6. What are the different steps in the downstream processing of biochemical products?
7. Outline the downstream processing steps in citric acid manufacture.
8. Discuss the steps involved in the product isolation and purification of an enzyme.

CHAPTER

2

Cell Disruption

2.1 INTRACELLULAR PRODUCTS

In bioprocesses, a wide variety of cells of procaryotic bacteria, eucaryotic fungi, plants and animals are used as hosts in the production of a diverse range of products. Some of the products are secreted into the medium but many are retained within the host cell. Examples of intracellular products formed by natural producers include glucose isomerase, phosphatase, β-galactosidase, ethanol dehydrogenase, *D*nase and *R*nase, $NADH/NAD^+$ and alkaloids. The advent of recombinant DNA (*r*DNA) technology has facilitated the production of a variety of metabolic products. In most of the genetically engineered species, the proteins produced remain within the cells. Many of the intracellular products are proteins and other products which are of complex structure and are fragile with respect to their biological activity. Hence the need for efficient methods for the release of intracellular products. Examples of such products include chymosin (yeast/*E.coli*), insulin (mammalian/*E.coli*), immunoglobulin, plasminogen activator, interferons (mammalian), human growth hormone (*E.coli*), human serum albumin, somatostatin, F VIII (mammalian) and streptokinase (mammalian). Cell disruption becomes a necessary pretreatment in the recovery of such products. It is natural that knowledge of the structure of the cell wall of the various organisms is required in selecting cell disruption techniques.

2.2 CELL WALL

The Gram positive bacteria such as *Lactobacillus* and *Streptococcus* have the simplest cell wall structure consisting of two layers (molecular weight exclusion ~1200) as shown in Figure 2.1(a). The outer layer is the murein layer, about 10–80 nm thick, made of peptidoglycan which exists in one form or another in almost all the species. The space below the murein

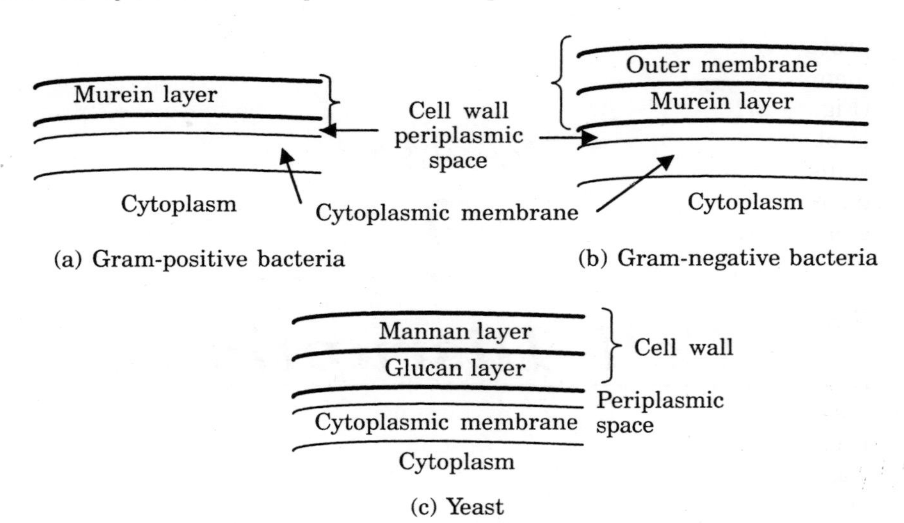

Figure 2.1 Structure of cell wall of microorganisms.

layer called periplasmic space, is about 8 nm thick and often contains enzymes. The inner layer called the plasma or cytoplasmic membrane, about 8 nm in thickness, is a double layer made of phospholipids and some proteins and metal ions. The plasma membrane controls the permeability of nutrients into the cell and metabolites from the cell into the surrounding solution.

The basic cell envelope of Gram negative bacteria such as *E.coli* and *Pseudomonas* sp. consists of three layers (molecular weight exclusion ~700), as shown in Figure 2.1(b). The outer membrane, about 8 nm thick, is made of a polymer containing both protein and lipopolysaccharide. A second (thinner) layer about 3 nm thick, is called murein layer. These two outer layers provide mechanical strength and it is necessary to rupture these membranes to release the intracellular products. Below it the periplasmic space exists. The third layer is the cytoplasmic membrane, similar in composition and function to the cytoplasmic membrane in Gram positive bacteria.

The cell interior, called the cytoplasm, is an aqueous solution of salts, sugars, amino acids and biopolymers including proteins, enzymes, RNA and DNA.

In naturally occurring procaryotes, the proteins are in solution whereas in genetically engineered species, excess protein is synthesized and precipitated within the cytoplasm. It is necessary to rupture the cell to release the protein in the cytoplasm.

Eucaryotic cells have nuclei and the cell wall is structurally more complicated. The yeast cell wall is of 70 nm thickness and consists of two main layers (molecular weight exclusion ~700). The outer layer is a highly crosslinked structural layer made of mannan (polysaccharide) and protein and below it lies the crosslinked glucan layer. The cytoplasmic membrane beneath the cell wall is a double layer made of phospholipids (Figure 2.1(c)). The precise cell wall structure of fungi is not known. It is of diverse nature, but mostly made of cellulose and chitin. Plant cell walls are mostly made of cellulose and other polysaccharides. Animal cells do not have a cell wall, but the cytoplasmic membrane is stiffened by sterols.

2.3 CELL DISRUPTION

Any potential method of cell disruption must ensure that the labile materials are not denatured by the process or hydrolyzed by the enzymes present in the cell. The various methods of cell disruption may be classified broadly into three types: (1) physical methods, (2) chemical and enzymatic methods and (3) mechanical methods. Some of the methods are essentially laboratory scale methods and only a few are used in the industrial scale.

2.3.1 Physical Methods

These include (i) heat shock or thermolysis, (ii) osmotic shock and (iii) ultrasonication.

Thermolysis. This is relatively an easy and economical method but can be used only if the products are stable to heat shock. Basically it inactivates the organisms by disrupting the cell wall without affecting the products. The effect of heat shock depends on parameters such as pH, ionic strength, presence of chelating or sequestering agents such as ethylene diamine tetraacetic acid (EDTA) which binds magnesium (thereby destabilizing the cell wall) and presence of proteolytic and other hydrolytic enzymes.

Osmotic shock. Osmotic shock to the cells is provided by simply dumping a given volume of cells into pure water of about twice the volume of cells. The cells swell due to osmotic flow of water ultimately bursting, thereby releasing the products into the surrounding medium. The osmotic pressure, π, of the cytoplasmic solution inside the cell, which causes the osmotic flow, is proportional to the concentration of the solutes and temperature, as given by van't Hoff equation.

$$\pi = RTC \tag{2.1}$$

where π refers to the difference in pressure inside the cell, P (in) being greater than atmospheric pressure, P (out). R is the gas constant, T is absolute temperature and C (moles per litre) is the total concentration of all the solutes in the cell. The susceptibility of the cells to undergo disruption by osmotic shock depends on their type. Red blood cells are easily disrupted. Animal cells can be disrupted only after mincing or homogenizing the tissues. Plant cells are more resistant to disruption by this method. Osmotic shock effect is often minimal on bacterial cells. However, the method is useful particularly if the desired products (enzymes) are located in the periplasmic region. Most of the bacteria have cell constituent concentration equivalent to 0.1–0.2 M NaCl solution. The osmotic pressure developed may be calculated using equation 2.1. For example, assuming the concentration of the cell constituents to be equivalent to 0.1 M NaCl, the osmotic pressure within the cell at 25 °C will be $0.1 \times 2 \times 0.082 \times 298 = 4.88$ atm. Many Gram positive bacteria have internal osmotic pressures in the range of 20 atm, but the cell integrity is maintained by a rigid mucopeptide layer and hence the osmotic method is not suitable for such bacteria. The method has been used for releasing hydrolytic enzymes and membrane binding proteins from a number of Gram negative bacteria including *E.coli* and *Salmonella typhimurium*.

Ultrasonication. This is essentially a laboratory method as it is expensive. Ultrasound waves of frequencies greater than 20 kHz rupture the cell walls by a phenomenon known as cavitation. The passage of ultrasound waves in a liquid medium creates alternating areas of compression and rarefaction which change rapidly. The cavities formed in the areas of rarefaction rapidly collapse as the area changes to one of compression. The bubbles produced in the cavities are compressed to several thousand atmospheres. The collapse of bubbles creates shock waves which disrupt the cell walls in the surrounding region. The efficiency of the method depends on various factors such as the biological condition of the cells (age and maturity), pH, temperature, ionic strength and time of exposure. Ultrasonication leads to a rapid increase in the temperature and to avoid heat denaturation of the product it is necessary to cool the medium and also limit the time of exposure.

2.3.2 Chemical and Enzymatic Methods

These include: (i) alkali treatment, (ii) detergent solubilization, (iii) lipid solubilization or cell wall permeabilization by organic solvents and (iv) enzymatic method.

Alkali treatment. This is a cheap and effective method but very harsh. Alkali acts on the cell wall in a number of ways including saponification of the lipids. Alkali treatment is carried out at pH 11–12 for about 20–30 minutes. Proteases are inactivated by this treatment and hence it is possible that the method would be useful in the preparation of pyrogen-free therapeutic enzymes (pyrogens are cell wall fragments of Gram negative bacteria or lipopolysaccharides and when ingested into mammals cause rise in body temperature). The enzyme l-asparaginase has been isolated by this method.

Detergent solubilization. This method involves the addition of a concentrated solution of detergent to about half the solution's volume of cells to disrupt the cell wall. The process depends on pH and temperature. Detergents are amphipathic capable of interacting with both water and lipid. The key mechanism involves the action of detergents in solubilizing the lipids in the cell wall to form micelle. In dilute solutions, the detergents do not dissolve but at higher concentrations lipid solubilization begins suddenly and thereafter increases linearly with detergent concentration. The surface tension of the medium shows abrupt change at the same concentration (see, Figure 2.2). The range of detergent concentration at which the abrupt changes in lipid solubility and surface tension of the medium occur is called *critical micelle concentration* and corresponds to the formation of micelle.

Examples of detergents used in cell wall disruption include anionic detergents such as sodium dodecyl sulphate (SDS), sodium sulphonate and sodium taurocholate, cationic detergents such as cetyltrimethyl ammonium bromide (CTAB) and non-ionic detergents such as Triton *X*-100. Sodium taurocholate is the natural detergent present in the bile salt of all animals except lobster which has alkyl sulphonate. Bile salts are known to increase the solubility of cholesterol by about 2 million times from 20×10^{-6} to 40 g/l.

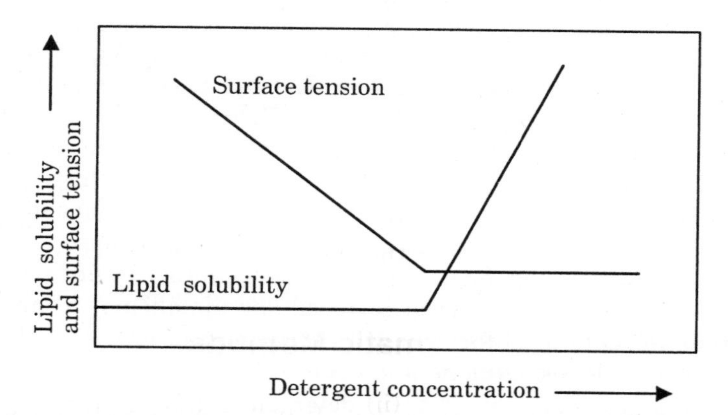

Figure 2.2 Effect of detergent concentration on lipid solubility and surface tension of the suspension.

In the case of mammalian cells, a non-ionic detergent such as saponin or steroid glycoside digitonin (or digitin), which binds β-hydroxysterols and is capable of complexing membrane cholesterol has been used to release intracellular proteins by permeabilizing the plasma membrane alone without affecting the organalle membranes.

Cell wall permeabilization. Cell wall disruption is achieved by the addition of organic solvents. For example, addition of toluene (10% of the cell volume), brings about cell wall disruption. The solvent is absorbed by the cell wall resulting in its swelling and ultimate rupture. Organic solvents at lower concentrations (1–3%) permeabilize the cell wall without disrupting it as shown in Figure 2.3.

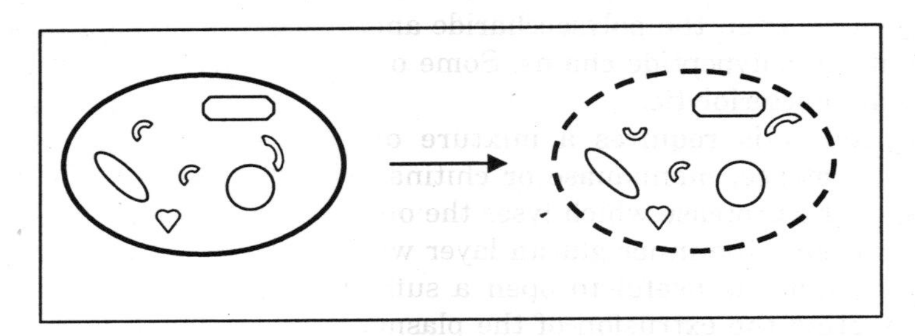

Figure 2.3 Schematic diagram of cell wall permeabilization.

This method is useful in retaining the contents of the cell for sequential release of desired products and also using the permeabilized cells as a porous bag of catalysts. Toluene permeabilization of cell wall has been practiced in the case of *Agrobacterium radiobacter* which produces two enzymes hydantoinase and *N*-carbamylamido hydrolase useful in the enzymatic conversion of hydantoin and substituted hydantoins to optically pure *D*-phenyl glycine and *D*-*p*-hydroxyphenylglycine required for the manufacture of semisynthetic penicillins. The enzymes are retained within the permeabilized cells and the substrate diffuses through the pores into the cell and the products diffuse out into the medium. Toluene permeabilization of yeast cells has been reported to release histidyl-*t*-RNA synthetase and enolase. Ethyl acetate has been used to permeabilize yeast cells to release periplasmic invertase and α-glucosidase. Dimethyl sulphoxide (DMSO) has been used to permeabilize plant cell walls (*Catharanthus rosens*) to release intracellular ajmalicine, a circulatory drug. Other solvents used for cell wall permeabilization include benzene, chlorobenzene, xylenes, cumene and octanol.

Enzyme digestion. Digestion of the cell wall is achieved by the addition of lytic enzymes to a cell suspension. Enzymes are highly selective, gentle and most effective, but costly. Only lysozyme has found commercial application in the industry, mainly for the extraction of enzymes, particularly glucose isomerase from *Streptomyces* sp. The enzyme hydrolyses α-1,4-glycosidic bonds in the mucopeptide moiety of bacterial cell wall of Gram positive bacteria. The final rupture of the cell wall often depends on the osmotic pressure of the suspending medium. In the case of Gram negative bacteria such as *E.coli*, pretreatment with a detergent such as Triton *X*–100, or addition of EDTA is necessary. EDTA is used to destabilize the outer membrane thereby making the peptidoglycan layer accessible to lysozyme. It is possible to form *E.coli* spheroplasts, cells with intact inner membrane, but porous or completely removed outer membrane by the combined use of lysozyme and EDTA.

In addition to lysozyme, three other types of bacteriolytic enzymes have been isolated. These include glycosidases, which split the polysaccharide chains, acetylmuramyl-*L*-alanineamidases, which cleave the junction between the polysaccharide and peptide and endopeptidases, which split the polypeptide chains. Some of the proteases also have been shown to be bacteriolytic.

Yeast cell lysis requires a mixture of different enzymes such as glucanase, protease, mannanase or chitinase. A mixture of two types of enzymes, a lytic protease which lyses the outer mannoprotein layer of the cell wall exposing the inner glucan layer which is acted on by the second enzyme glucanase is useful to open a sufficiently large hole in the cell wall, facilitating the extrusion of the plasma membrane and its contents intact as protoplast in osmotic support buffers containing 0.55 to 1.2 M sucrose or mannitol. In dilute buffers, the protoplast also lyses releasing the cytoplasmic proteins and organelles which may also lyse. Plant cells can be lysed by cellulase and pectinase.

The development of expression systems for recombinant proteins and recombinant protein particles located in specific cell locations necessitates the use of highly selective techniques for cell lysis and product release. Mechanical methods are not specific and tend to release the desired product along with a host of other biomolecules and cell debris. Enzyme lysis of microbial cells has advantages in that controlled sequential degradation of cell wall can give rise to selective product release thereby preventing their contamination and denaturation.

Sequential disruption of microbial cells for selective product release involves the use of lytic enzymes depending on parameters such as the type of microbial cells, products to be harvested and process conditions. Microbial cells, particularly yeast cells engineered for producing exclusively cell wall polymers, human serum albumin or hepatitis B surface antigen in conjunction with special lytic enzyme systems have been reported. The schematics of the sequential disruption of microbial cells for selective product release are shown in Figure 2.4.

The method involves the removal of the extracellular products followed by the addition of lytic enzymes to the microbial cells suspended in an osmotic support medium to disrupt the cell wall. Cell wall disruption results in the release of cell wall proteins leaving the protoplast intact. In the second step, the protoplast is disrupted by gentle agitation. The soluble cytoplasmic enzymes are separated from recombinant proteins and mitochondria. The organalle products are released in the final step by the use of appropriate enzyme or reagent. For example, yeast cells have been disrupted in a sequential two step process to release invertase and glucan polymer without affecting the protein particle. In the first step, cell wall digestion by protease released invertase in the absence of any osmotic support medium and in the second step, the use of glucanase enzyme released glucan polymer.

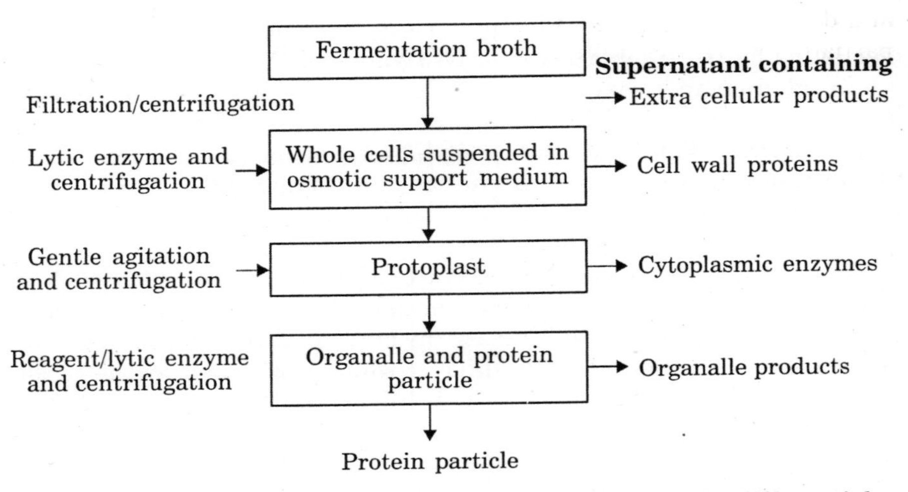

Figure 2.4 Sequential disruption of microbial cell and differential release of intracellular products.

Potential industrial applications of the sequential disruption method include (i) preparation of protoplasts, cell fusion and transformation of yeast, (ii) production of intracellular enzymes, (iii) pretreatment to increase yeast digestibility, (iv) preparation of soluble glucan polysaccharide, (v) alkali extraction of yeast protein, (vi) extraction of specialized lipids from yeast, production of yeast extract, food preservation, (vii) extraction of pigments from red yeast, (viii) release of recombinant proteins (e.g., human serum albumin), (ix) ethanol recovery from spent brewer's yeast and (x) pretreatment of microbial cells for mechanical rupture. Various bacteriolytic and yeast lytic enzymes are produced by a number of microorganisms such as *B.subtilis, Staphylococcus, Streptomyces globisporius, Cytophaga, Micromonospora* and *Arthrobacter*. The yeast cell wall degrading enzyme kitalase from the fungus *Rhizoctonia* sp. has been used in the liberation of cell-bound invertase from *S.cerevisiae.*

2.3.3 Mechanical Disruption of Cells

Mechanical methods find use in the laboratory as well as in industrial scale operations. In the laboratory, wet grinding with a ball mill or a Waring blender may be used. Waring blender is particularly effective with animal cells and tissues as well as with mycelial organisms. In industrial scale, cell disruption is carried out by using a (i) bead mill or (ii) high pressure homogenizer.

Bead mill disruption. The vertical or horizontal bead mill consists of a grinding cylinder with a central shaft fitted with a number of impellers and driven by a motor as shown in Figure 2.5. The grinding cylinder is partially filled with grinding elements or beads made from wear resistant

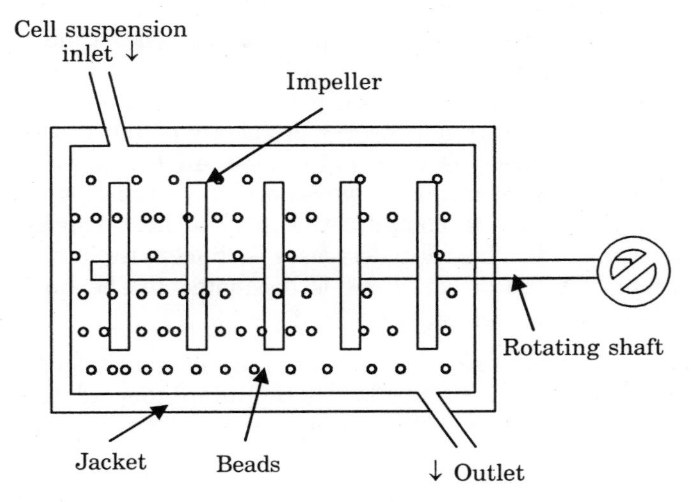

Figure 2.5 Schematic diagram of a bead mill.

materials such as glass, alumina, titanium carbide, zirconium oxide or zirconium silicate. Horizontal bead mills have the advantages of higher loading of beads (up to 80%) of smaller size, uniform distribution of beads for good grinding at lower speeds and lower energy input. The cell suspension is pumped into the cylinder and cell disruption occurs due to shear forces produced between velocity gradients because of the rotary movement of cells and beads. In addition, collision between beads and cells and grinding of cells between rolling beads also contribute to the disruptive forces. The rate and degree of cell disruption depend on several parameters. These include (i) the nature of microorganism—its size, cell wall composition and thickness and concentration of microbial cells, (ii) product location within the cell—cytoplasm, periplasmic region or organalle, (iii) type of bead mill—type of impeller, its agitation speed (specifically the tip speed of the impeller), (iv) bead size, its density and loading, (v) residence time and (vi) temperature. The effect of these parameters on the rate and extent of product release need to be determined experimentally.

Cell disruption and release of the product in a bead mill may be described by first order kinetics. Assuming that the concentration of the product that can be released from a given amount of cell suspension is C_{max}, the concentration of the product C released at a given time, t, may be given by the first order rate Eq. (2.2).

$$\ln\left(\frac{C_{max}}{C_{max} - C}\right) = -kt \tag{2.2}$$

where k is the first order constant. The value of k depends on the type of impeller, bead size and loading, speed of agitation and temperature. It has to be determined experimentally. Equation 2.2 is valid for batch mode operation of a bead mill with a specified residence time. In continuous mode of operation, the mean residence time and the residence time distribution or the number of CSTR units in series need to be considered for calculating the rate and degree of release of the product. Scaling up the bead mill process is difficult because all the power input via the impeller is dissipated into heat which has to be removed via the cylinder wall. It is necessary to compute the ratio of heat transfer area to the volume of the mill. This ratio decreases with increase in cylinder diameter with which, all the power input is dissipated in the broth, raising the temperature of the cell suspension.

High pressure homogenizer. A high pressure homogenizer consists of a high pressure positive displacement pump coupled to an adjustable discharge valve with a restricted orifice as shown in Figure 2.6.

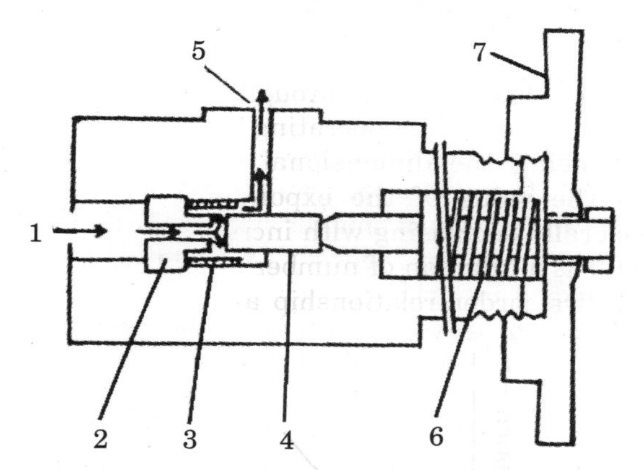

Figure 2.6 Schematic diagram of a high pressure homogenizer valve assembly. (1) feed (2) valve seat (3) impact ring (4) valve (5) discharge (6) spring assembly and (7) pressure control wheel.

The cell suspension is pumped through the homogenizing valve at 200–1000 atmospheric pressure depending on the type of microorganism and concentration of the cell suspension. The disrupted cell suspension is cooled, as it exits the valve, to minimize thermal denaturation of sensitive products. Cell disruption occurs due to various stresses developed in the fluid. The primary mechanism seems to be the stress developed due to the impingement of the high velocity jet of suspended cells on the stationary surface (impact ring). The stress developed is expressed as dynamic pressure P_s which depends on the jet velocity v and fluid density ρ as given by Eq. (2.3):

$$P_s = \frac{1}{2}\rho v^2 \tag{2.3}$$

In addition, other stresses are generated by the homogenizer including normal and shear stresses. The normal stress is generated as the fluid passes through the narrow channel of the orifice and the shear stress as the pressure rapidly reduces to almost atmospheric pressure as the cell suspension passes out of the orifice.

Different parameters influence the degree of cell disruption and the rate of release of the product. These include (i) the nature of microorganism—its size, cell wall composition and thickness, and concentration of microbial cells, (ii) product location within the cell—cytoplasm, periplasmic region or organalle, (iii) type of homogenizer—type of valve and seat, (iv) operating pressure, (v) temperature and (vi) number of passes of the cell suspension through the homogenizer. The optimal conditions need to be evaluated experimentally.

Cell disruption in homogenizer may also be described on the basis of first order kinetic expression as given by Eq. (2.4).

$$\ln\left(\frac{C_{\max}}{C_{\max} - C}\right) = -kN \tag{2.4}$$

where N is the number of passes through the valve and k is the first order rate constant depending on operating pressure (and hence) given by $k = k'P^n$, (k' being the dimensional rate constant, P the operating pressure and the value of the exponent n varying over a range of pressures, generally decreasing with increasing pressure). The disruption of baker's yeast as a function of number of passages at different operating pressures is a first order relationship as shown in Figure 2.7.

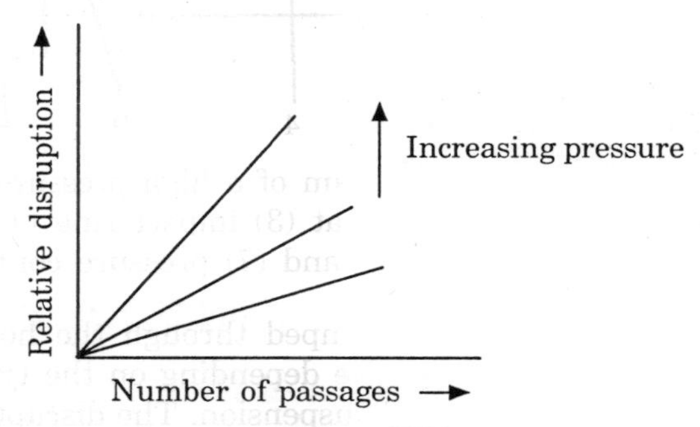

Figure 2.7 Disruption of yeast cells at different operating pressures.

Scale up of homogenizers is relatively simple in that only a bigger plunger pump and discharge valve are required provided all other variables are maintained constant. Another advantage in the scale up of a homogenizer is that the increase in temperature is a function of only the operating pressure and is independent of the size of the homogenizer.

Release of intracellular enzymes by mechanical disruption. The release of intracellular enzymes is more complex compared to the release of proteins because the enzymes may be located in different regions within the cell. For example, the cytoplasmic enzyme glucose-6-phosphate dehydrogenase is released along with proteins from baker's yeast while fumarase present in the mitochondria is released slowly by mechanical disruption as shown in Figure 2.8. In contrast, other enzymes such as acid phosphatase, invertase, α-glucosidase and restriction endonuclease EcoR I which are located in the periplasmic region outside the cytoplasmic membrane are released faster than cytoplasmic proteins. Thus, the rate of enzyme release can be an indication of its location within the cell. The rate of release of the enzyme and the rate of inactivation of the enzyme need to be monitored as a function of time in a batch process to get the maximum yield of the desired product. Enzymes that are less stable show a maximum in such a plot; for example, alcohol dehydrogenase (see Figure 2.8) indicating that the enzyme inactivation occurs due to shear stress, prolonged residence time or increase in temperature.

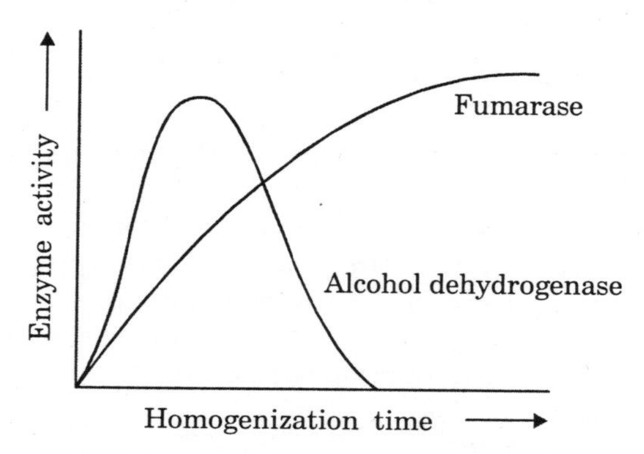

Figure 2.8 Release of different enzymes by mechanical disruption.
Source: P.A. Belter, E.L. Cussler and Wei-Shou Hu, *Bioseparations—Downstream Processing for Biotechnology*, John Wiley & Sons, New York, 1988.

2.4 PROTEINS OF INCLUSION BODIES

Expression of eucaryotic proteins in *E.coli* through recombinant DNA technology often results in the aggregation of the expressed proteins and the formation of insoluble protein inclusion bodies. The recovery of the proteins in their active form is a problem to be tackled in downstream processing. A two step procedure involving (i) dissolution of the protein body and (ii) protein folding is adopted to achieve the desired result.

Dissolution of the protein body. In the first step, the inclusion bodies are released from the host cells by lysis and low speed centrifugation to remove cell debris. The inclusion bodies are washed and separated from other impurities by high speed centrifugation. The inclusion body pellets are solubilized by chaotropic reagents such as 6 M guanidine hydrochloride or 8 M urea or by anionic detergents such as sodium dodecyl sulphate in the presence of reducing agents such as β-mercaptoethanol to break the disulphide bonds in the proteins. The dissolution of proteins occurs due to denaturation and simultaneous dissociation of the subunits of the proteins. The chaotropic reagents alter the structure of water, thereby weakening the hydrophobic forces holding the proteins in aggregates. Detergents bind to the proteins and mask the hydrophobic regions of the proteins facilitating their dissolution. The dissolved proteins may then be separated/purified by usual methods.

Protein folding. The denatured protein has to undergo correct folding through the formation of disulphide bonds to become active. The removal of denaturing agents from the solution alone is not sufficient for recovery of the native protein. One of the several methods developed for reforming the native protein involves protecting the free cysteine residues during purification of the denatured protein subunits by reversible blocking of the sulphhydryl groups. Using this method, the native human insulin with correct intra-chain disulphide bonds was obtained from the two peptide subunits of the human insulin expressed as inclusion bodies in different strains of *E.coli*. In another method, the refolding of the denatured protein has been achieved by the use of the enzyme thioredoxin. The enzyme facilitates a quick 'search' through various pairs of cysteine residues to arrive at the correct combination of disulphide bridges by catalyzing disulphide interchange until the free-energy minimum of the native protein is achieved. The method has been demonstrated in refolding the completely denatured or scrambled RNase.

Questions

1. Give a description of the cell wall structure of microbial cells.
2. Discuss in detail the physical methods of cell disruption.
3. Outline the mechanism of action of a detergent on the cell wall.
4. What is cell permeabilization? What is its use?
5. Give an account of the action of enzymes in cell disruption. How is the method useful in sequential release of products?
6. Write notes on the operation and functioning of (a) bead mill and (b) homogenizer in cell disruption.
7. What are protein inclusion bodies? How are they converted to native proteins?

CHAPTER

3

Filtration

3.1 SOLID–LIQUID SEPARATION

The products of bioprocesses in the laboratory or industry are mostly encountered as suspended solids or dissolved solutes necessitating the adoption of different separation steps to isolate them from fermentation broths or beers. The concentration of the particulates may vary from as low as 0.1% to as high as 60% (w/v) depending on the process. Further the size of the particles range between 1 μm diameter for microorganisms to about 1 mm diameter for nutrients. The preliminary step in bioseparations would always involve the removal of suspended particles consisting of cell debris to clarify the beer to make it amenable for further separation steps to isolate the desired product. Filtration is the conventional unit operation aimed at the separation of particulate matter.

Filtration is defined as the separation of solid in a slurry consisting of the solid and fluid by passing the slurry through a septum called the filter medium. The slurry is pumped usually at a pressure in perpendicular direction to the filter medium as shown in Figure 3.1.

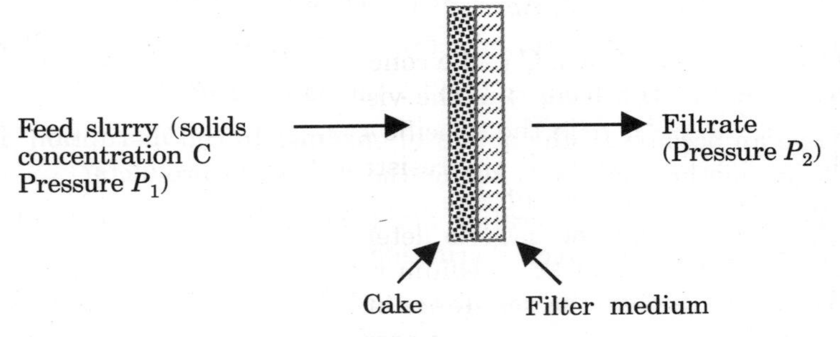

Figure 3.1 Schematic diagram of filtration.

The filter medium allows the fluid (filtrate) to pass through and retains the solids. The separated solid called the filter cake forms a bed of particles on the filter medium. The thickness of the cake increases from an initial value of zero to a final thickness at the end of filtration. During filtration, the filtrate therefore passes first through the cake and then the filter medium. The fluid experiences a pressure drop ($\Delta p = P_1 - P_2$) as it flows through the filter cake. The pressure drop is the sum of the pressure drop across the filter cake (Δp_c) and across the filter medium (Δp_m).

Filtration may be carried out such that Δp is constant (constant pressure filtration) wherein the volume rate of filtrate flow dV/dt will decrease as filtration progresses. Alternatively, filtration may be carried out such that the rate of filtration $1/A\ (dV/dt)$, is constant (constant volume filtration) by increasing the Δp as filtration progresses.

3.2 THEORY OF BATCH FILTRATION

The rate of filtration is defined as the volume of filtrate V (in m^3) collected per unit time, t (in minutes) per unit area A (in m^2) of the filter medium and is expressed as:

$$\text{Rate of filtration} = \frac{1}{A}\frac{dV}{dt} \tag{3.1}$$

where (dV/dt) refers to the volume rate of filtrate flow. The flow of the filtrate through the cake is through the void space between the solid particles forming the bed of cake and the flow is generally laminar. Under these conditions the working equation for filtration is

$$\frac{dt}{dV} = \frac{\eta}{A\,\Delta p}\left\{\frac{VC\alpha}{A} + r_m\right\} \tag{3.2}$$

or

$$\frac{dt}{dV} = KV + B \tag{3.3}$$

where

$$K = \frac{\eta C\alpha}{A^2\Delta p} \qquad \text{and} \qquad B = \frac{\eta r_m}{A\Delta p}$$

In the above equations, C is the concentration of solids (kg m^{-3}) in the total volume V of the feed, η is the viscosity of the filtrate (in Pa s or Nm^{-2} s or kg m^{-1} s^{-1}), α is the specific resistance of the cake in m kg^{-1} (length per mass units), r_m is the resistance of the filter medium and Δp is the pressure drop of the filtrate.

Equation (3.3) may be used to determine the values of α and r_m. In general, the resistance of the medium r_m is constant and is independent of the cake. In contrast, the resistance of the cake varies with the amount of feed volume V linearly for incompressible cakes and non-linearly for compressible cakes. In the case of incompressible cakes, the cake thickness is directly proportional to the filtrate total volume and inversely proportional to the area of the filter medium.

Integrating Eq. (3.3) between the limits $t = 0$, $V = 0$ to $t = t$ and $V = V$, the time required for filtering a broth of volume V (also the filtrate volume V) under constant pressure is given as

$$t = \frac{KV^2}{2} + BV$$

or

$$\frac{t}{V} = \frac{KV}{2} + B \tag{3.4}$$

A plot of t/V versus V would be linear as shown in Figure 3.2, with the slope depending on the pressure drop and properties of the cake, represented by its specific resistance and concentration in the feed. The slope describes the increasing resistance of the cake as filtration progresses. The intercept is directly proportional to the resistance of the medium and is independent of the cake.

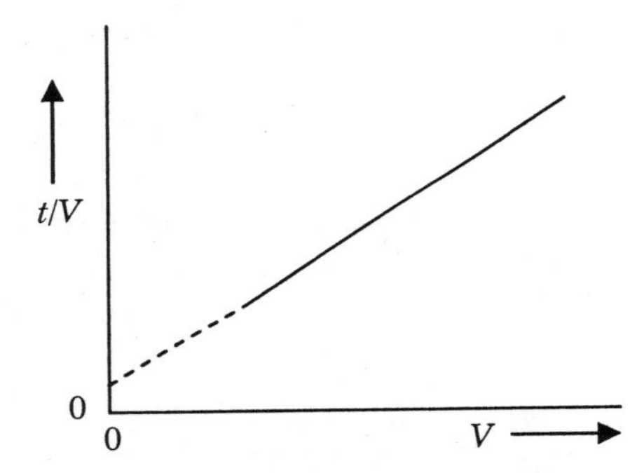

Figure 3.2 Constant pressure filtration.

The resistance of the medium is negligible in most of the cases as indicated by the intercept being very close to zero. The Eq. (3.4) then simplifies with the second term (BV) becoming zero.

3.2.1 Batch Filtration Involving Compressible Cakes

Some filter cakes change shape as the cake builds up during filtration. Due to this, the voidage of the bed of particles in the cake changes. The bottom layers of the cake will be more compacted than the upper layers.

Generally, the fermentation beers are often difficult to filter mainly due to their high non-Newtonian viscosity and the formation of highly compressible cakes. The cakes formed during filtration also deform into impermeable mat, particularly with mycelial microorganisms. Various factors affect the specific cake resistance in broth filtration. These include the type of microorganism, fermentation time, pH and temperature of the broth and size distribution of particles. For example, the effect of fermentation time on the specific cake resistance of *Penicillium chrysogenum* broth is shown in Figure 3.3.

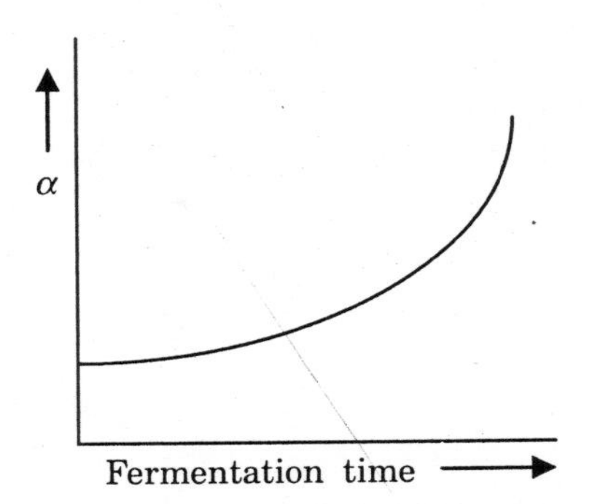

Figure 3.3 Variation of cake resistance of *Penicillium chrysogenum* fermentation broth as a function of time.

Source: J. Krijgsman, *Product Recovery in Bioprocess Technology*, BIOTOL series, Butterworth-Heinemann Ltd., Oxford, 1992.

The specific cake resistance varies by a factor of 10 or more because of the fact that the mycelium will be more fragmented as fermentation time increases. The effects of these factors on specific cake resistance have to be determined experimentally in order to optimize the conditions of filtration process.

In the case of compressible cakes usually formed in most of bioseparations, the drop in filtration rate as filtration progresses is not strictly linear because the slope as described by Eq. (3.4) is a complex function of pressure drop. However, the intercept characterizing the resistance of the medium is not affected by the characteristics of the cake. Under these circumstances, the effect of the compressibility of the cake on its resistance is assumed to be a function of pressure drop as given by,

$$\alpha = \alpha'(\Delta p)^s \quad \text{or} \quad \log \alpha = \log \alpha' + s \log \Delta p \tag{3.5}$$

where α' is a constant related to the size and shape of the cake particles and s is cake compressibility, its value varying from zero for a rigid incompressible cake to close to 1 for a highly compressible cake. In practice, the value of s varies between 0.1 and 0.8. The compressibility of the filter cake can be determined experimentally by conducting filtration experiments at different Δp values. If the α values obtained for different Δp values remain independent of Δp, the filter cake is said to be incompressible. However, if the α values vary with Δp, the cake is said to be compressible.

A graphical plot of log α vs. log (Δp) (Figure 3.4) based on equation 3.5 would give the values of the intercept log α' and the slope s. When the value of s is high, pretreatment of the feed with filter aids

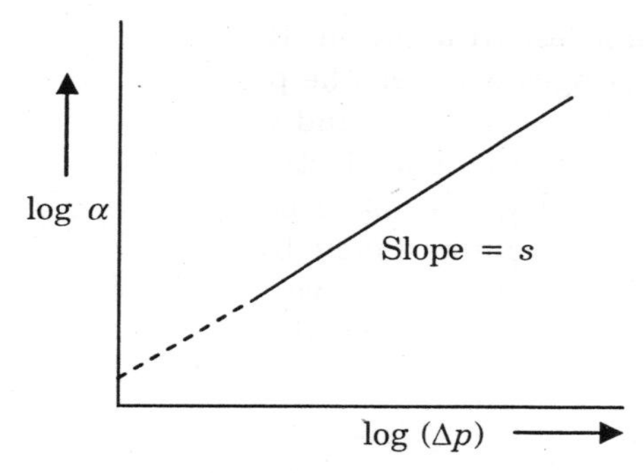

Figure 3.4 Effect of pressure drop on cake resistance.

becomes necessary. For compressible cakes the time of filtration is calculated by substituting $(\Delta p)^{1-s}$ for Δp into Eq. (3.4).

3.3 PRETREATMENT OF FERMENTATION BROTHS

The separation of suspended solids in fermentation broths is facilitated by subjecting the broths to pretreatment to enhance their filterability. Any one of the following three general methods may be adopted depending on requirements. The available methods of pretreatment include (i) heating, (ii) coagulation and flocculation and (iii) adsorption on filter aids.

3.3.1 Heating

This method is the simplest and the least expensive method provided the volume is small. It improves the broth's handling characteristics and has the advantage of pasteurising the broth. For example, on heating a dilute solution of egg white (containing ovalbumin), the protein is irreversibly denatured, but the threads of cooked egg are much easier to filter than the original solution. However, the method is not suitable for thermally labile products.

3.3.2 Coagulation and Flocculation

Addition of chemicals such as acids, bases, simple electrolytes or polyelectrolytes promote coagulation and flocculation of the broth. Acids and bases change the pH and hence the charge on the particles, facilitating coagulation and settling of solids.

Simple electrolytes or flocculants such as ferric chloride and aluminum sulphate (alum) screen the electrostatic repulsion between colloidal particles thereby allowing the van der Waals' and London attractive forces to operate and coagulate the colloidal particles to dense and large sized particles. In addition, they act as buffers and also form inorganic polyoxo bridges between the particles leading to agglomeration of particles. These chemicals are found to be more effective when used in conjunction with synthetic polyelectrolytes. Polyelectrolytes such as polyacrylics and polyamines are also used in the range of 0.1–2%. Cationic flocculants such as polyamines may be preferentially used for broths containing cells because the cells have a net negative charge at neutral pH. Flocculants, however, have the disadvantage of fouling membrane filter presses.

3.3.3 Adsorption on Filter Aids

Filter aids are inert, incompressible discrete particles of high permeability. Solids such as wood pulp, starch powder, cellulose, inactive carbon, diatomaceous earth (skeletal remains of tiny aquatic plants deposited centuries earlier) and perlites (volcanic rock processed to yield an expanded form) are added as filter aids. The addition of a filter aid or body feed in 0.5 to 5% (w/w) concentration to fermentation broths enhances their filterability. In the case of mycelial broths, small particles of fragmented mycelium or bacterial cells penetrate into the pores of the filter medium or precoat thereby reducing their permeability. Filter aids adsorb these small particles preventing the clogging of filter medium. Filter aids also reduce the compressibility of the accumulated biomass by adsorbing the colloidal particles thereby decreasing the specific cake resistance as shown in Figure 3.5. The particle size of filter aids vary between 2 and 20 μm. However, with large sized particles of certain filter aids, the clarity of filtrate is less. Another disadvantage is that certain antibiotics may bind irreversibly to filter aids as in the case of

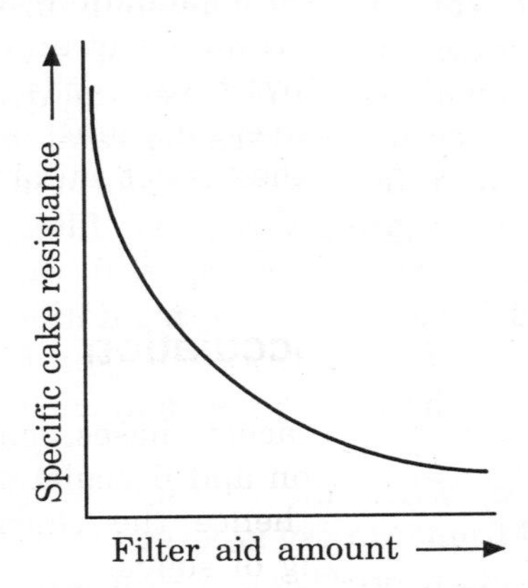

Figure 3.5 Effect of filter aid on the specific resistance of a compressible cake.

aminoglycosidic antibiotics to diatomaceous earth. The type of filter aid to be used for a particular broth and its concentration need to be optimized by preliminary studies.

3.4 FILTER MEDIA

The two main functions of the filter medium are to promote cake formation and to mechanically support the cake once it is formed. The chosen medium should offer minimum resistance to the flow of the liquid consistent with the rapid formation of cake. The medium should be strong enough to support the cake under operating conditions. The medium should be non-toxic and chemically inert towards the material being filtered. The surface characteristics of the medium should facilitate easy removal of cake.

Rigid media such as sand, gravel, diatomaceous earth or charcoal find use in water treatment and in certain bioseparations but rarely in food industry. Porous carbon, porcelain, alumina, perforated metal sheets and wire meshes find use in a variety of industries. Flexible filter media such as woven fabrics of cotton, silk, wool and jute and synthetic membranes of nylon, polythene, polypropylene and terylene as well as non-woven materials such as cotton or wool fibers and paper pulp are also used. The filter medium is chosen with specific pore size and to suit particular filtration application on the basis of preliminary small scale trials.

3.5 EQUIPMENT

For batch filtration, filter press, horizontal or vertical plate and frame filter press, pressure leaf filter are commonly used.

The *plate and frame press* consists of several plates and frames arranged alternately on a horizontal or vertical frame and held in position by means of a hand screw or hydraulic ram so that there is no leak between the plates and frames which form a series of liquid tight compartments. The plates are covered with filter media made of cloth or membrane. The filter press is the cheapest filtration equipment, requires least floor space and is suitable for fermentation broths with low solids content and low resistance to filtration (see Figure 3.6). A dry cake discharge is obtained. The filter press may also be used to collect high value solids.

The *pressure leaf filter* incorporates a number of leaves, each consisting of a metal framework of grooved plates covered with a fine wire mesh or filter cloth often precoated with a layer of cellulose fibers (see Figure 3.7). The feed slurry is fed into the filter which is operated under pressure or by suction with a vacuum pump. The equipment is amenable for sterilization with steam. It is suitable for large volumes of liquids with low solid content and for small batch filtration of high value solids.

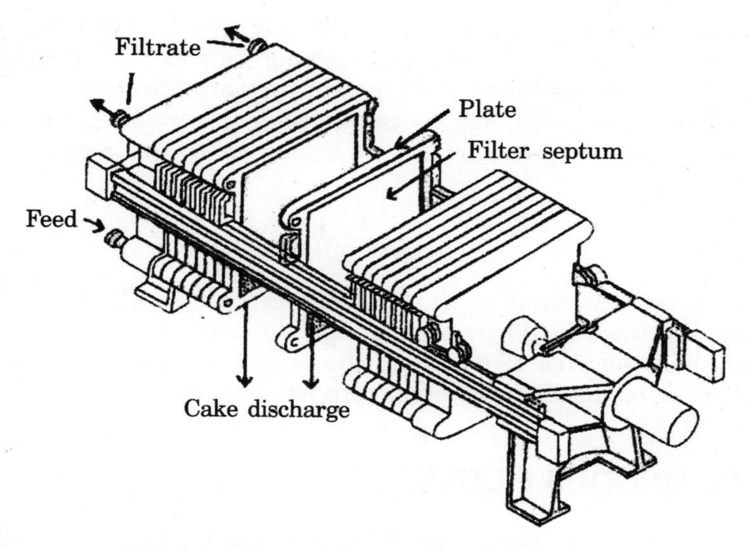

Figure 3.6 Plate and frame filter press.

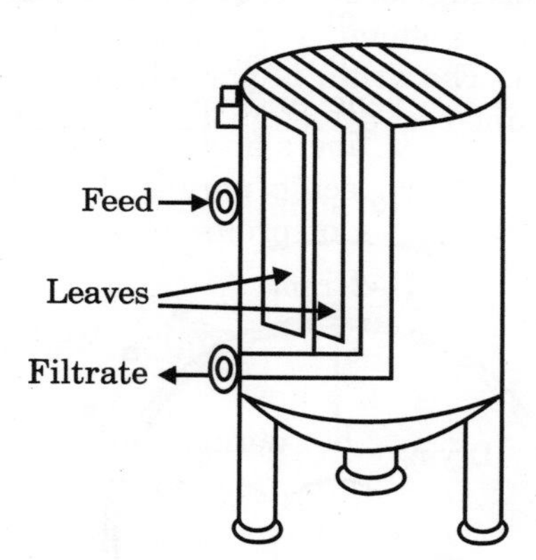

Figure 3.7 Pressure leaf filter.

Source: P.A. Belter, E.L. Cussler and Wei-Shou Hu, *Bioseparations—Downstream Processing for Biotechnology*, John Wiley & Sons, New York, 1988.

3.6 WASHING OF FILTER CAKES

The total filtration time in a batch filtration involves the washing and cleaning cycles apart from the time required for filtering a given volume of feed as given by Eq. (3.4). The washing of the filter cake after filtration involves the displacement of the filtrate by the wash liquid and by diffusion of the filtrate. The amount of wash liquid should be sufficient to displace all the filtrate from the cake. The washing rate is calculated by assuming that the cake structure is not affected during washing and the conditions during washing remain the same as those at the end of filtration cycle.

In batch filtration using plate and frame filter press or leaf filter, the final filtering rate gives the predicted washing rate as the wash liquid follows the flow path of filtrate. For constant pressure filtration using the same pressure in washing step as in filtering, the washing rate in a pressure leaf filter is given by Eq. (3.6), which is the reciprocal of Eq. (3.3).

$$\frac{dV}{dt} = \frac{1}{KV + B} \tag{3.6}$$

where (dV/dt) is the washing rate in m^3/s. However, in the plate and frame filter press, the wash liquid travels through the cake twice as thick and across only half the area of filtration, the predicted washing rate being only one-fourth of the final filtration rate as given by Eq. (3.6).

3.7 CONTINUOUS FILTRATION

Rotary vacuum filter or *rotary drum filter* is widely used in industries for continuous filtration involving large volumes of feed. The equipment consists of a rotating hollow segmented perforated drum covered with a fabric or metal filter mesh. The drum is partially submerged in the feed slurry to be filtered contained in a trough as shown in Figure 3.8.

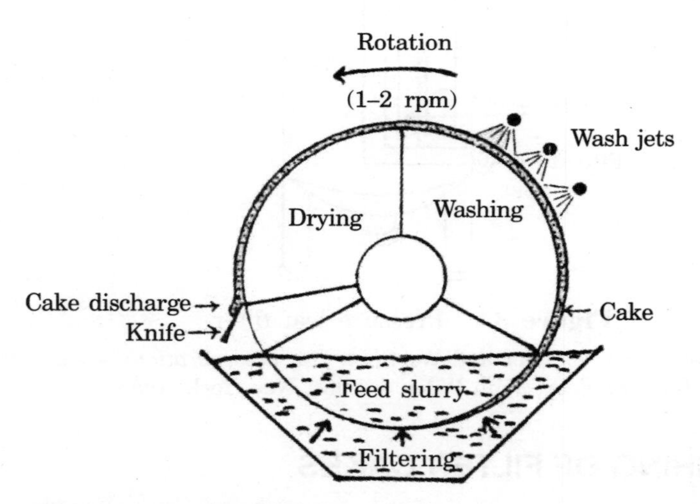

Figure 3.8 Schematic diagram of a rotary vacuum filter.

The interior of the drum is divided into a series of compartments to which vacuum is applied. As the drum revolves slowly at about 1 rpm, the liquid is sucked through the filter medium covering the outer surface of the drum and the solids are retained on the outer surface forming a cake. The liquid passes through the interior of the drum to the collection vessel. The cake formed adhering to the drum comes out of the feed trough as the drum rotates. It is washed with water from spray jets, dried and removed by a knife or doctor blade at the end of the rotation of the drum. In some filters, a string discharge mechanism assisted by the release of vacuum and a small blow back or a continuous belt discharge mechanism in which the filter medium moves over a series of rollers operate to remove the cake are used.

The filter medium of the rotary drum filter is usually precoated with a thick layer of diatomaceous earth or an inert microporous material. The inert precoat material may also be added to the feed slurry. During filtration, the filter aid and biomass accumulate on the surface of the precoat and continue to build in the cake formation segment of the filtration cycle. The cake is washed and dewatered. A slowly advancing doctor blade shaves off a thin layer of accumulated biomass along with the filter aid and a part of precoat, thereby exposing a fresh surface of precoat for the next filtration cycle. Thus a dynamic filter with longer effective lifetime is obtained. However, there is a disadvantage in that the volume of the solids increases.

The operation cycle of the continuous rotary drum filter consists of three steps: (i) cake formation, (ii) cake washing to remove either valuable or unwanted solutes and (iii) cake discharge. The first two steps decide the filtration time and the second step decides washing efficiency and the volume of the wash liquid required.

The time required for the formation of cake, t_f, is given by the following equation.

$$t_f = \frac{KV^2}{2} + BV \tag{3.7}$$

The time t_f is related to cycle time t_c, which is the time required for one rotation as given by the relationship:

$$t_f = t_c\beta \tag{3.8}$$

where β is the fraction of time that the drum filter is submerged, or in other words, the fraction of cycle devoted to cake formation. Therefore Eq. (3.7) becomes:

$$t_c\beta = \frac{KV^2}{2} + BV \tag{3.9}$$

Substituting for K and B from Eq. (3.3),

$$t_c\beta = \frac{\eta C\alpha V^2}{A^2\Delta p} + \frac{\eta r_m V}{A\,\Delta p} \tag{3.10}$$

In continuous filtration the resistance of the filter medium is generally negligible compared to the cake resistance and hence the B term of Eq. (3.9) can be neglected. However when short cycle times are used or when the filter medium resistance is appreciable, Eq. (3.10) is more appropriate. For compressible cakes, the Δp term in Eq. (3.10) is substituted with $(\Delta p)^{1-s}$ to include the compressibility factor.

The flow rate of feed slurry or filtration flux in the rotary drum filter is given by Eq. (3.11).

$$\frac{V}{At_c} = \left(\frac{2\beta(\Delta p)^{1-s}}{t_c \eta C \alpha'}\right)^{1/2} \tag{3.11}$$

Two factors are involved in the second step of cake washing (1) the fraction of soluble material remaining after the wash, which decides the volume of wash liquid required and (2) the rate at which the wash liquid passes through the cake which decides the wash time. The fraction of solute remaining is related to the volume of wash liquid as:

$$r = (1 - \gamma)^n \tag{3.12}$$

where r is the ratio of soluble solutes remaining after the wash to that originally present in the cake prior to wash, n is the volume of wash liquid required divided by the volume of the liquid retained in the cake and γ is the wash efficiency of the cake. The value of r varies from zero to one, lower values of r corresponding to more effective washing. If $\gamma = 0$, then $r = 1$ and no reduction of soluble solutes in the cake is achieved whatever the volume of wash liquid used. A plot of log r against n gives a straight line whose slope is log $(1 - \gamma)$ as shown in Figure 3.9.

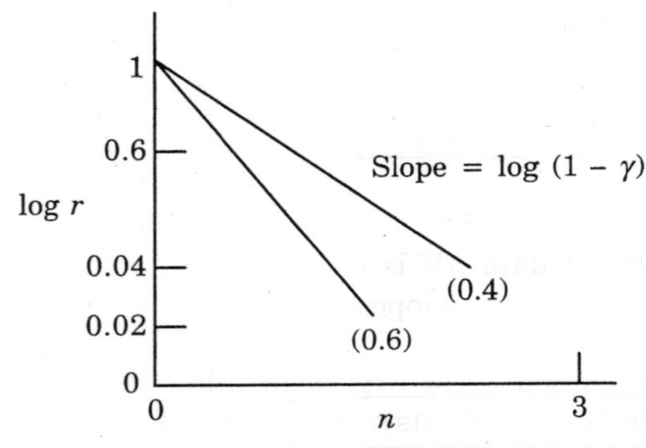

Figure 3.9 Plot of log r vs. n for different washing efficiencies.

Once the volume of liquid required to wash the cake is determined from the above equation, i.e., when the wash liquid contains no additional solids, the flow of the wash liquid will be constant and equal to the final instantaneous filtration at the end of the cake formation. The rate is given by

$$\frac{V_w}{A} = \left[\frac{(\Delta p)^{1-s}}{2\eta\alpha' C t_f}\right]^{1/2} t_w \tag{3.13}$$

where V_w and t_w are respectively the volume of wash water required and the time required for washing.

The third step, namely, cake discharge is instantaneous and does not contribute to the filtration time. As a post treatment, the discharged filter cake may be dried by hot air. The scale-up of rotary filtration depends on the resistance of the cake and the volume of cake.

Exercises

1. Calculate the specific resistance of the cake (α) and the resistance of the filter medium (r_m) on the basis of the following experimental data for a constant pressure filtration of a suspension of incompressible solids of concentration 30 kg/m^3 on a filter medium of area 1 m^2. The pressure drop was 2 bar. The viscosity of the filtrate (η) is 1.1×10^{-3} kg/m/s.

Vol. of filtrate V(m^3)	Time t (s)
2.0×10^{-3}	23
4.0	60
6.0	114
8.0	184
10.0	270

Solution

(i) For the given data t/V is calculated. From the straight line plot of t/V vs. V the slope = $K/2$ and the intercept = B are determined.

V (m^3)	t (s)	t/V (s m^{-3})
2.0×10^{-3}	23	11.5×10^3
4.0	60	15.0
6.0	114	19.0
8.0	184	23.0
10.0	270	27.0

From the graph, $K = 4.0 \times 10^3$ s m^{-6} and the intercept $B = 7.5$ s m^{-3}

(ii) From Eq. (3.3) the value of specific cake resistance and the resistance of the filter medium are calculated.

$$\alpha = \frac{KA^2 \Delta p}{\eta C}$$

$A = 1\ \text{m}^2$; $\Delta p = 2$ bar (1 bar $= 1 \times 10^5\ \text{Nm}^{-2}$ or $1 \times 10^5\ \text{kg m}^{-1}\ \text{s}^{-2}$)

$\eta = 1.1 \times 10^{-3}\ \text{kg m}^{-1}\text{s}^{-1}$; $C = 30\ \text{kg m}^{-3}$.

$$\alpha = \frac{(4.0 \times 10^3\ \text{s m}^{-6})(1\text{m}^4)(2 \times 10^5\ \text{kg m}^{-1}\text{s}^{-2})}{(1.1 \times 10^{-3}\ \text{kg m}^{-1}\text{s}^{-1})(30\ \text{kg m}^{-3})}$$

$$= 2.42 \times 10^{10}\ \text{m kg}^{-1}$$

and

$$r_m = \frac{BA\Delta p}{\eta}$$

$$= \frac{(7.5\ \text{s m}^{-3} \times 1\ \text{m}^2)(2 \times 10^5\ \text{kg m}^{-1}\text{s}^{-2})}{(1.1 \times 10^{-3}\ \text{kg m}^{-1}\text{s}^{-1})}$$

$$= 1.36 \times 10^9\ \text{m}^{-1}$$

2. A lab scale filtration experiment of a slurry with a solids content of 24.6 g/litre on a filter of area of 5 cm^2 at a pressure drop of 500 mm of Hg gave the following data. The filtrate had a viscosity of 1 cP.

Vol. of filtrate V (cm^3)	Time t (s)
30	6.0
40	10.0
50	16.0
60	23.0
70	31.0

(i) Calculate the time required for filtration of 2.0 m^3 of the same feed slurry on a plate and frame filter press consisting of 15 frames of area of 30 cm^2/frame.

(ii) Calculate the total filtration cycle time if the cake was washed with 0.25 m^3 of water and cleaning the press at the end of the filtration cycle requires 20 minutes.

Solution

(i) The values of α and r_m are obtained from the plot of t/V vs. V.

The t/V values for the above data are 0.2. 0.25, 0.32, 0.383 and 0.443 s cm^{-3}.

The intercept $B = 0$ and hence $r_m = 0$.

The value of $K = 1.4 \times 10^{-2}$ s cm^{-6}.

$\Delta p = 500$ mm Hg (1mm Hg at 0 °C $= 1.333 \times 10^2$ kg m s^{-2})

$$\alpha = \frac{(1.4 \times 10^{-2}\,\text{s cm}^{-6})(25\,\text{cm}^4)(6.665 \times 10^7\,\text{g cm s}^{-2})}{(0.01\,\text{g cm}^{-1}\text{s}^{-1})(24.6 \times 10^{-3}\,\text{g cm}^3)}$$

$$= 9.33 \times 10^{10}\ \text{cm g}^{-1}.$$

K is proportional to $1/A^2$ and B is proportional to $1/A$. The new value of K is calculated by substituting the given filter area A of 1.35 m^2 for the 15 frame filter press.

$$\text{The new } K \text{ value} = \frac{1.4 \times 10^{-2} \times 25}{(1.35 \times 10^4)^2} = 1.92 \times 10^{-9}\ \text{s cm}^{-6}$$

Substituting the new K value in Eq. (3.4),

$$t = \frac{(1.92 \times 10^{-9}\,\text{s cm}^{-6})(2.0 \times 10^6\,\text{cm}^3)^2}{2} = 3840.8\ \text{s} = 64\ \text{min.}$$

(ii) For calculating the rate of washing the cake on a plate and frame filter press, Eq. (3.6) may be written as:

$$\frac{dV}{dt}\,(\text{cm}^3\text{s}^{-1}) = \frac{1}{4} \times \frac{1}{KV + B}$$

$$= \frac{1}{4} \times \frac{1}{(1.92 \times 10^{-9}\,\text{s cm}^{-6})(2.0 \times 10^6\,\text{cm}^3) + 0}$$

$$= 65.1\ \text{cm}^3\,\text{s}^{-1}$$

At the above washing rate, the time required for filtering the wash liquor of volume 0.25 m^3 = 0.25×10^6 cm^3 = 3840 s = 64 min.

The total filtration cycle time = filtration time + washing time + cleaning time

= 64 + 64 + 20 = 148 min.

= **2.47 hours**

3. The specific resistance of the cake of biomass was found to vary with pressure drop as follows. Find the compressibility of the cake.

Pressure drop (kN/m^2)	Cake resistance (m/kg)
330	3.56×10^{11}
134.3	2.16×10^{11}
46.1	1.45×10^{11}
21.1	1.07×10^{11}

Solution

ln (α)	ln (Δp)
5.8	26.6
3.9	26.10
3.83	25.70
3.05	25.40

The plot of ln(α) versus ln(Δp) gives a straight line with slope s and intercept of ln α'.

S (compressibility) = 0.377; $\alpha' = 7.2 \times 10^{10}$ m/kg

4. A broth of 2 litres of viscosity 2.8 cP containing 5.2% of compressible biomass was filtered on a filter of 10 cm diameter at a pressure drop of 1.6 bar. The filtration time was 16 minutes. The cake was found to have a compressibility of 0.4. Calculate the time required to filter 2500 litres of the broth on filter press of area 2.5 m^2 at a pressure drop of 2.8 bar. Assume the resistance of the filter medium to be negligible.

Solution

The value of α is calculated from the given preliminary experimental data.

Time t = 16 min (960 s); V = 2 litres (2.0×10^{-3} m^3);
$A = \pi r^2 (r = d/2 = 5$ cm$) = 7.857 \times 10^{-3}$ m^2;
η = 2.8; cP = 2.8×10^{-3} kg/m/s;
C = 5.2% = 0.52 kg/ m^3; s = 0.4

The value of K is calculated by substituting the given data in Eq. (3.4) as 4.80×10^8 s cm^{-6} and that of α as 2.697×10^{10} m/kg.

The new value of K for the scale up involving a pressure drop of 2.8 bar and a filter area of 2.5 m^2 is 3387.5 s m^{-6}.

The total time to filter 2500 litres of broth on a filter of 2.5 m^2 at a pressure drop of 2.8 bar is calculated as 176 min. (2.93 hours).

Questions

1. Discuss the theoretical principles of constant pressure filtration.
2. How is compressibility of a cake determined?
3. How does the compressibility of the cake alter the filtration characteristics of the broth?
4. Write a note on the necessity of pretreatment of fermentation broths and the methods available for pretreatment.
5. Give an account on the working of a continuous rotary filter.
6. How is filtration time determined for continuous filtration of a broth?
7. How is washing efficiency determined in rotary drum filtration?

CHAPTER

4

Centrifugation

4.1 CENTRIFUGAL FORCE

Centrifugation is used to separate materials of different densities when gravitational force is insufficient for separation. In industry centrifugal force is used (i) to separate fine solids from liquids by centrifugal sedimentation, (ii) to separate immiscible liquids whose density difference is small by centrifugal decantation and (iii) in the filtration of solids from liquids by centrifugal filtration. It is advantageous to opt for centrifugation, particularly when filtration becomes a cumbersome operation due to small size of the particles. However, centrifugation produces a paste of solids and more often, a concentrated suspension, which requires further dewatering, as against the dry cake obtained by filtration. The equipment is costlier compared to that required for filtration. The equations relevant to the centrifugal force acting on a particle during centrifugation are summarized below.

During the circular motion of a centrifuge, the centrifugal force F_c (in Newton, N) acting on a particle is related to the angular velocity ω (rad/s) and the radial distance, r (m), of the particle from the centre of rotation as:

$$F_c = ma = m\omega^2 r \tag{4.1}$$

where $\omega^2 r$ = acceleration due to centrifugal force. The tangential velocity of the particle, v (m/s) is given by the relationship, $v = \omega r$. Equation (4.1) may be rewritten as:

$$F_c = mr\left(\frac{v}{r}\right)^2 = \frac{mv^2}{r} \tag{4.2}$$

The rotational speed of a centrifuge (n) is generally expressed in terms of number of revolutions per minute (rpm). The centrifugal force may be expressed in terms of number of revolutions of the centrifuge by substituting the angular velocity in Eq. (4.1) by the relationship $\omega = 2\pi n/60$.

$$F_c = mr\left(\frac{2\pi n}{60}\right)^2 = 0.01097\ mrn^2 \tag{4.3}$$

The centrifugal force may also be expressed in terms of gravitational force F_g (= mg where g is the acceleration due to gravity, 9.80665 m/s^2) as

$$\frac{F_c}{F_g} = \frac{\omega^2 r}{g} = \frac{v^2}{rg} \tag{4.4}$$

$$= \frac{r}{g}\left(\frac{2\pi n}{60}\right)^2 = 0.001118\ rn^2\ (\text{since } n = 60v/2\pi r) \tag{4.5}$$

Thus the force developed in a centrifuge is $\omega^2 r/g$ or v^2/rg times as large as the gravity force and is often expressed as equivalent to so many g forces.

4.2 CENTRIFUGAL SEDIMENTATION

The basic principle involved in centrifugal separation of solids is the density difference between the solids and the surrounding fluid. Normally a suspension of solids in a liquid on standing settles down slowly under the influence of gravity. The process is known as sedimentation. In centrifugation, the process of settling is aided by centrifugal forces. When a solid particle moves through a viscous medium (infinite continuum), its velocity is affected by two opposing forces, the gravitational and drag forces. The particle is accelerated by the gravitational force resulting from the density difference between the particle and the surrounding fluid. The gravitational force, F_g, acting on spherical particles (a good theoretical approximation for many biological particles) is quantitatively given by Eq. (4.6).

$$F_g = \frac{\pi}{6}[d^3(\rho_s - \rho)]\,g \tag{4.6}$$

where d is the sphere's diameter, ρ_s and ρ are the densities of the sphere and the fluid respectively and g is gravity. Equation 4.6 is obtained by combining the buoyancy force (Eq. (4.6a)) and the gravitational force (Eq. (4.6b)) acting on the particle.

$$F_b = \frac{\pi}{6}[d^3\rho]\,g \tag{4.6a}$$

$$F_g = \frac{\pi}{6}[d^3\rho_s]\,g \tag{4.6b}$$

Equation (4.6) parallel's Newton's definition of force ($F = ma$) because the term in the square brackets is the effective mass of the sphere.

The drag force, F_d, acting on a single spherical particle in solution is given by Stoke's law as,

$$F_d = 3\,\pi d \eta v \tag{4.7}$$

where η is the viscosity of the medium (Pa s or N s m^{-2} or kg m^{-1} s^{-1}) and v is the velocity of the spherical particle (m s^{-1}). The equation holds good only when the sphere is small so that Reynolds number, characterizing the flow around the sphere, given as $(dv\rho/\eta)$ is less than 1. This condition is almost always satisfied for biological solutes as the particle size is very small.

Initially as the spherical particle begins to move in the solution, the drag force is small, as the velocity of the particle is small. The particle accelerates and reaches its terminal velocity when the buoyancy and drag forces are counterbalanced by gravitational force (i.e., $F_g = F_b + F_d$). The steady state terminal velocity v_g of the spherical particle under such conditions is obtained by combining Eq. (4.6) and Eq. (4.7) as

$$v_g = \frac{d^2}{18\eta}(\rho_s - \rho)\,g \tag{4.8}$$

In settling process, the acceleration is due to gravity only and the force acting on the spherical particle is represented by g.

Sedimentation in centrifugal field is similar but the force acting on the spherical particle is centrifugal force, and hence the constant g is replaced by $\omega^2 r$ (see Figure 4.1). The terminal velocity v_c in centrifugal separation of solids is expressed in terms of angular velocity, ω, in radians/s and the radial distance, r, from the centre of the centrifuge to the spherical particle in cm. The terminal velocity of the particle under centrifugal force is then given by:

$$v_c = \frac{d^2}{18\eta}(\rho_s - \rho)\,\omega^2 r \tag{4.9}$$

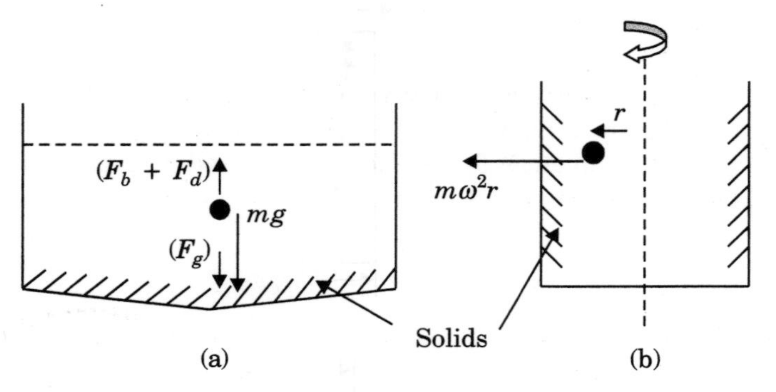

Figure 4.1 (a) Gravity sedimentation and (b) centrifugal sedimentation.

4.3 CENTRIFUGAL DECANTATION

Gravity decantation is ineffective when the densities of two immiscible liquids are not very different and hence centrifugal decantation needs to be adopted. When a mixture of light (lower density) and heavy (higher density) liquids is fed into a centrifuge rotating at a high speed, the heavy liquid moves towards the wall of the centrifuge bowl and the light liquid towards the centre of the bowl thereby getting separated. Centrifugal decantation finds use in the dairy industry to separate cream from milk.

4.4 CENTRIFUGES

The common types of centrifuges that find application in bioseparations include: (i) tubular bowl centrifuge, (ii) multichamber bowl centrifuge, (iii) disc stack centrifuge with or without nozzle, (iv) decanter centrifuge and (v) basket centrifuge.

4.4.1 The Tubular Bowl Centrifuge

It is the simplest type and can provide very high centrifugal force. The centrifuge can be cooled and hence is advantageous in protein and other thermally labile bioproduct separation. The centrifuge is mostly used in pilot plant level. It consists of a long narrow cylindrical bowl suspended from the top rotating at high speed of about 10,000 rpm, in an outer stationary casing. Bowl dimensions range from 8 to 15 cm in diameter and up to 150 cm in height. The schematic diagram of the tubular bowl centrifuge is shown in Figure 4.2.

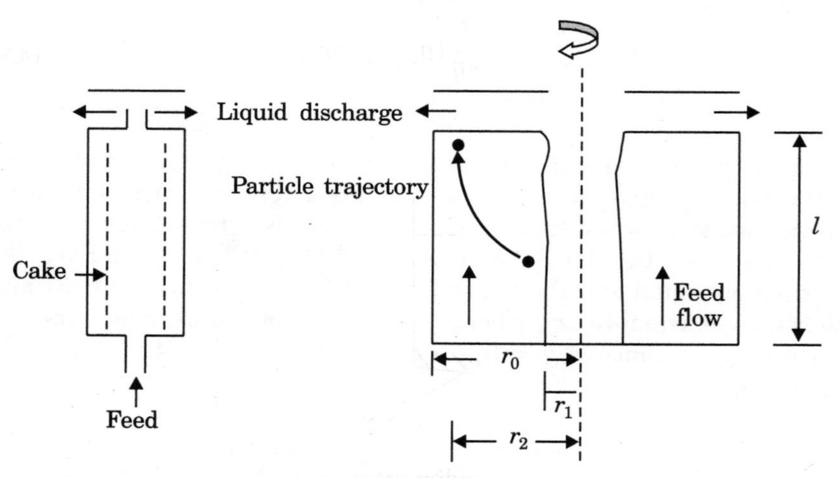

Figure 4.2 Tubular bowl centrifuge and flow pattern in the centrifuge.

The feed is introduced at the bottom of the bowl and the discharge of the supernatant occurs through an annular opening at the top. The feed liquid moves upward at a uniform velocity carrying with it the solid particles. The solids deposit on the bowl's inner wall as a thick paste. The suspension can be fed until the solid loss in the effluent becomes prohibitive, upon which the bowl must be dismantled and cleaned.

The trajectory of the particle and the distances from the axis of rotation to the wall of the centrifuge (r_0), to the liquid interface (r_1) and to the particle (r_2) are shown in Figure 4.2. At the end of the residence time in the centrifuge, if $r_2 \approx r_0$, then the particle deposits on the wall of the centrifuge and if $r_2 < r_0$, the particle leaves the bowl along with the fluid. The particle is assumed to move radially at a terminal velocity given by equation 4.9. Since $v_c = dr/dt$, Eq. (4.9) becomes,

$$dt = \frac{18\eta}{\omega^2(\rho_s - \rho)d^2}\frac{dr}{r} \tag{4.10}$$

Integrating between the limits $r = r_1$ at $t = 0$ and $r = r_0$ at $t = t_r$ (residence time)

$$t_r = \frac{18\eta}{\omega^2(\rho_s - \rho)d^2}\ln\left(\frac{r_0}{r_1}\right) \tag{4.11}$$

Since $t_r = V/Q$, where V is the volume (m^3) of the bowl and Q is feed flow rate or the volumetric capacity (m^3/s), Q is given by

$$Q = \frac{\omega^2(\rho_s - \rho)d^2}{18\eta\ln\left(\frac{r_0}{r_1}\right)}V \tag{4.12}$$

The volume of the tubular bowl is given by $V = \pi l(r_0^2 - r_1^2)$ where l is the length of the bowl. Substituting for V in Eq. 4.12 we obtain:

$$Q = \frac{\omega^2(\rho_s - \rho)d^2}{18\eta\ln\left(\frac{r_0}{r_1}\right)}[\pi l(r_0^2 - r_1^2)] \tag{4.13}$$

Only those particles having diameters given by Eq. (4.13) and larger particles will be deposited on the wall at the given flow rate and within the residence time while smaller sized particles will be moving with the liquid. In most tubular centrifuges, the distances r_0 and r_1 are approximately equal and the factor $[(r_0^2 - r_1^2)/\ln(r_0/r_1)]$ in Eq. (4.13) may be shown to be equal to $2r^2$, where r is an average radius, approximately equal to r_0 or r_1. Since $v_g/g = [(\rho_s - \rho)d^2/18\eta]$ from Eq. (4.8), Eq. (4.13) may be simplified as:

$$Q = v_g\frac{[2\pi lr^2\omega^2]}{g} \tag{4.14a}$$

$$= v_g[\Sigma] \tag{4.14b}$$

The maximum possible feed flow or volumetric capacity of the tubular bowl centrifuge, Q, is thus related to the terminal velocity of the particle v_g and the different parameters of the centrifuge represented by the factor Σ. The factor Σ may be defined as the equivalent clarification area with dimensions of (length)^2. The term indicates the required area of a gravity-settling tank with the same clarifying characteristics as the centrifuge for the same feed rate. The importance of Eq. (4.14b) is that it implies the terminal velocity v_g is a function only of the particles and is independent of the type of centrifuge that and the centrifuge parameters represented as Σ is a function of only of the particular centrifuge as it depends on centrifugal design and is independent of the particles. Hence the factor Σ is useful for comparing the different types of centrifuges and in the scale-up of centrifuges. For example, to scale up from a lab centrifuge with capacity Q_1 and characteristics Σ_1 to a large scale Q_2 the relationship to be used is $Q_2 = \Sigma_2(Q_1/\Sigma_1)$, i.e., the dimensions of the large scale centrifuge required to achieve the same terminal velocity as in the lab scale version is arrived at.

4.4.2 The Multichamber Bowl Centrifuge

It contains a number of concentric tubes connected in such a way that a zigzag flow of the feed suspension through the chamber is achieved (see Figure 4.3).

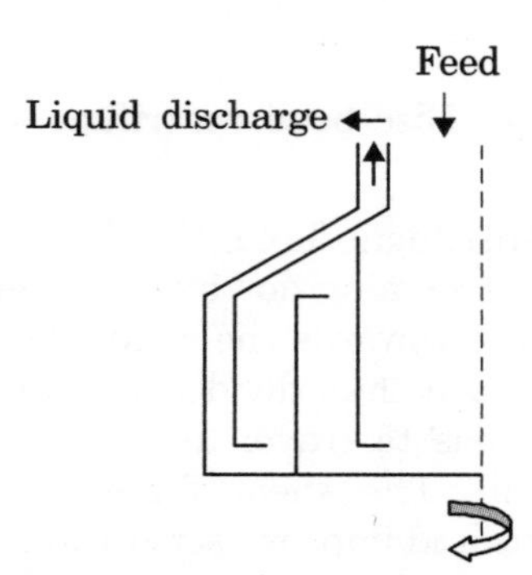

Figure 4.3 Multichamber bowl centrifuge.

The solids are deposited on the outermost chamber wall and solid discharge is done manually. Multichamber bowl centrifuge is employed in the fractionation of human blood plasma.

4.4.3 The Disc (Stack) Bowl Centrifuge

It consists of a shallow wide cylindrical bottom driven bowl rotating at moderate speed (about 6000 rpm) in a stationary casing. The bowl, about

30 to 100 cm in diameter contains a number of closely spaced metal discs, located one above the other with a fixed clearance of about 0.5–2 mm between them obtained by placing spacer bars. The discs have one or more set of matching holes, which form a channel through which the feed material flows, as the discs rotate with the bowl. The feed is introduced at the bottom through a centrally located feed pipe from above and the clarified liquid flows out through an annular slit near the neck of the bowl as shown in Figure 4.4.

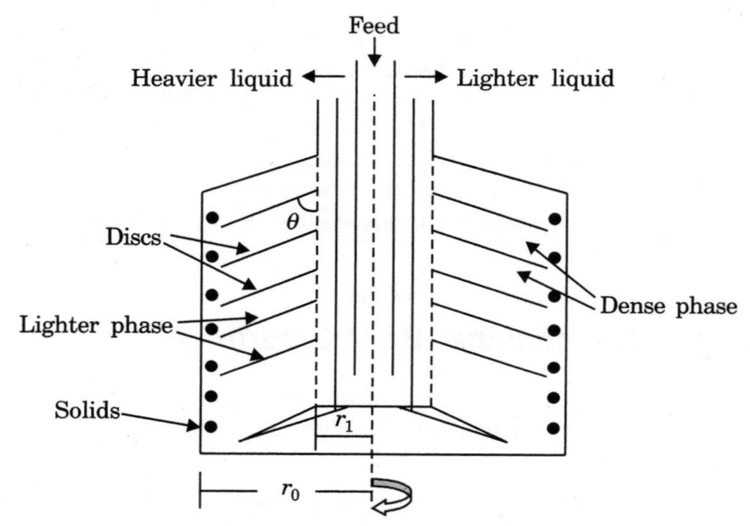

Figure 4.4 Disc bowl centrifuge.

Under the influence of centrifugal force, the dense phase of the feed travels towards the bowl wall streaming down the undersides of the discs while the lighter phase displaced towards the centre flows over the upper faces of the discs. The liquids are thus divided into thin layers and the distance any drop of a liquid has to travel to get in to the appropriate outgoing stream is very small. The shear forces at the liquid–liquid interface break down emulsions and improve separation of the phases. The solids may be removed intermittently or continuously from the sides. The discharged solids have relatively a higher amount of liquid compared to the discharge from the tubular centrifuge.

The feed flow Q, is given by considering the number of discs n, the angular rotation ω, the angle θ at which the discs are fitted from the vertical and the distances r_0 and r_1 of the outer edge and inner edge respectively of the discs from the axis of rotation as given by Eq. (4.15).

$$Q = v_g \frac{[2\pi n\omega^2]}{3g} (r_0^3 - r_1^3) \cot\theta \tag{4.15}$$

$$= v_g[\Sigma]$$

Disc bowl centrifuge is used in starch-gluten separation and cream separation from milk.

4.4.4 The Decanter or Scroll Type Centrifuge

This is mainly used to concentrate slurries with high dry solid concentrations. The decanter consists of a rotating horizontal bowl, with a length:diameter ratio of 1:4, fitted with a screw conveyor and an adjustable feed pipe located in the middle of the decanter. The screw rotates slightly faster than the bowl. The solid particles deposited on the wall are scraped off while the liquid leaves the machine via an overflow weir.

4.5 SELECTION OF CENTRIFUGE TYPE

The selection of the type of centrifuge to be used depends on the physical properties of the slurry and the relative centrifugal force (RCF or G-number). The physical properties of the slurry include mainly the volume of solids and the particle size. The multichamber bowl centrifuge can handle only dilute suspensions containing 0.1 to 10% solids (v/v) content in the particle size range of 0.5–500 μm. The decanter is best suited for larger sized particles (5–10,000 μm) and higher concentration of solids in the range of 10–60% of solids content.

The relative centrifugal force of the rotor is defined as given by the equation obtained by comparing Eq. (4.8) and (4.9).

$$G = \frac{\omega^2 r}{g} \tag{4.16}$$

where r is the radius of the machine (in m), ω is the angular speed in rad/s and g is the gravitational constant (m/s^2). The RCF factors commonly used for laboratory centrifuges are in the range of 5,000 for small table centrifuges. This means the settling velocity of the particle in the centrifuge will be 5,000 times the settling velocity of the same particle in gravity settling. Hence the time of separation is small in a centrifuge revolving at a high angular velocity. The RCF factors are 50,000 for a high speed centrifuge and 500,000 for an ultracentrifuge. The RCF values for industrial centrifuges are smaller compared to laboratory centrifuges due to the slower speed of rotation caused by the limitations of stress in the bowl. Thus the RCF factor for a tubular centrifuge is the range of 13,000–17,000, while for the disc stack centrifuge the range is 5,000–13,000 and for the decanter, 1,500–4,500 which are much smaller.

4.6 SCALE-UP

Scale-up and selection of centrifuges for large scale operations is based on any one of the two approaches, namely, (i) equivalent time also called the *Gt*-method or (ii) [Σ] factor method.

The *Gt*-method is a qualitative one in which, by calculating the product of centrifugal force G and time t, the difficulty of a given separation is assessed.

$$Gt = \frac{\omega^2 r}{g} t \tag{4.17}$$

where r is the radius of the centrifuge.

The *Gt* values are determined using a laboratory scale centrifuge called gyro tester. The gyro tester has graduated test tubes and the RCF or G for given speeds is calibrated and known. The graduated test tube is filled with about 10 ml of suspension and centrifuged for 5–30 seconds at fixed speeds and the *Gt* values are calculated. The *Gt* values for different types of biomass vary widely, from as low as 0.3×10^6 seconds for eucaryotic cells and chloroplasts to as high as 1100×10^6 seconds for lysosomes and ribosomes. Other typical *Gt* values are 2×10^6 seconds for eucaryotic cell debris and cell nuclei; 9×10^6 seconds for protein precipitates; 18×10^6 seconds for bacteria; and 54×10^6 seconds for mitochondria and bacterial cell debris. A large scale centrifuge with *Gt* value that is comparable to the laboratory determined *Gt* value for a given bioseparation is selected.

The Σ factor method is based on the relationship expressed by Eq. (4.6). The gyro tester is useful for determining the velocity v of the particles in a given bioseparation. The Σ factor for tubular bowl centrifuge and disc bowl centrifuge has been given by Eq. (4.14b) and (4.15) respectively. A large scale centrifuge is chosen for a given separation based on the process requirements of v and Q.

4.7 CENTRIFUGAL FILTRATION

Centrifugal filtration is adopted to separate solids from liquid, in place of pressure filtration. Filtration time is usually short when centrifugal force is used, for a given volume of filtrate to be collected. The *basket centrifuge* involves a combination of centrifuge and a filter and hence the process is called centrifugal filtration. The basket centrifuge consists of a perforated cylindrical basket, which rotates rapidly. The suspension is fed along the axis of the bowl and solids accumulate on the wall of the basket. The liquid flows out under centrifugal forces through the cake and the perforations in the basket wall (see Figure 4.5). The centrifuge is useful for washing the accumulated solids.

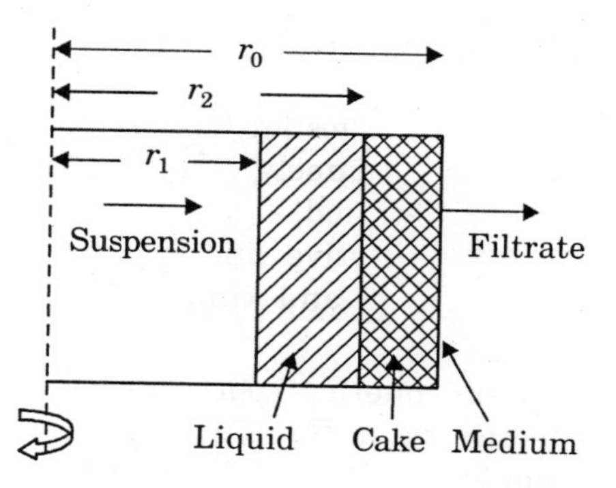

Figure 4.5 Schematic diagram of a basket centrifuge.

Centrifugal filtration can be considered as similar to normal filtration involving the flow of the liquid through a bed of solids (cake). In normal filtration where a flat cake is formed, the pressure drop Δp is proportional to the liquid velocity, v as given by Eq. (4.18).

$$\frac{\Delta p}{h} = v(\eta \alpha \rho_0) \tag{4.18}$$

where $h(= r_0 - r_2)$ is the thickness of the cake, η is the liquid viscosity, α is the specific cake resistance and ρ_0 is the density of solids per volume of liquid. However in centrifugal filtration, the cake is not flat and the pressure drop varies with the radius. Hence the above equation is replaced by Eq. (4.19).

$$-\frac{dp}{dr} = v(\eta \alpha \rho_0) \tag{4.19}$$

Further the velocity of the liquid through the cake is not constant and is higher at the centre of the centrifuge. Thus the total volumetric flow, Q and the thickness of the cake vary with time; Q decreasing with increasing thickness of the cake. The entire pressure drop is essentially due to the centrifugal force on the fluid given as $\Delta p = 1/2\rho\omega^2(r_0^2 - r_1^2)$ (ρ is the density of the liquid) and Q is given by Eq. (4.20).

$$Q = \frac{\pi \omega^2 \rho l (r_0^2 - r_1^2)}{\eta \alpha \rho_0 \ln(r_0/r_2)} \tag{4.20}$$

The volume filtered per unit area (V/A) in centrifugal filtration is given by Eq. (4.21).

$$\frac{V}{A} = \frac{\rho_c}{\rho_0}(r_0 - r_2) \tag{4.21}$$

where ρ_c and ρ_0 are the density of solids per unit volume of cake and per unit volume of liquid respectively and r_0 and r_2 are as shown in Figure 4.5 and $(r_0 - r_2)$ is the thickness of the flat cake.

The time required for centrifugal filtration, t is given by Eq. (4.22).

$$t = \frac{\eta \alpha \rho_c^2}{2\rho_0 \Delta p}(r_0 - r_2)^2 \qquad (4.22)$$

Exercises

1. A suspension of spherical particles of 0.1 mm diameter was allowed to settle in a column of 50 cm length. The density difference between the solid particles and the liquid was 0.05 g/cm^3 and the viscosity of the liquid was 1.1 cP. (i) Calculate the settling time of the particles assuming that the particles reach their terminal velocity almost instantaneously. (ii) Calculate the settling time in a centrifuge rotating at 400 rpm if the distance between the axis of rotation and bottom of the centrifuge was 12 cm and the distance between the axis and the liquid surface was 3 cm.

Solution

(i) Substituting the given data into Eq. (4.8), the terminal velocity of the particles, v_g is calculated.

$$v_g = \frac{(0.01\,\text{cm})^2\,(0.05\,\text{g cm}^{-3})(980\text{ cm s}^{-2})}{(18 \times 0.011\,\text{g cm}^{-1}\,\text{s}^{-1})}$$

$$= 0.0247\text{ cm/s}$$

The time for settling is given by $t = l/v_g$, where l is the length of the column.

$$t = 50\text{ cm}/(0.0247\text{ cm s}^{-1}) = 2024.3\text{ s or }33.7\text{ min.}$$

(ii) From equation 4.11,

$$\ln\left(\frac{12\,\text{cm}}{3\,\text{cm}}\right) = \frac{(0.01\,\text{cm})^2}{18 \times 0.011\,\text{g cm}^{-1}\,\text{s}^{-1}}(0.05\,\text{g cm}^{-3})\left[\frac{400 \times 2 \times (22/7)}{60\,\text{s}}\right]^2 \times t$$

$$1.3863 = 0.04434\,t$$

$$t = 31.26\text{ s}$$

2. The centrifugal separation of a biomass of 80 μm sized cells of density 1.04 kg m^{-3} was carried out in a tubular centrifuge having a diameter of 15 cm and rotating at 1200 rev/min. (i) Calculate the residence time if the distance between the liquid surface and the axis of rotation was 0.8 cm, the liquid density and the liquid viscosity were 1.0 kg m^{-3} and 0.013 g cm^{-1} s^{-1} respectively

(ii) What would be the volumetric capacity of the centrifuge if its length was 40 cm? (iii) Calculate the Σ factor (iv) What will be the time required for centrifuging 1000 litres of broth?

Solution

(i) The residence time t_r is calculated by substituting the given data into Eq. (4.11).

$$t_r = \frac{18(0.013\text{ g cm}^{-1}\text{ s}^{-1})[\ln(7.5\text{ cm}/0.8\text{ cm})]}{(1200 \times 2\pi/60\text{ s})^2(1.04 - 1.0\text{ kg m}^{-3}(80 \times 10^{-6}\text{ m})^2}$$

$$= \frac{0.5237\text{ g cm}^{-1}\text{s}^{-1}}{0.040458\text{ g cm}^{-1}\text{s}^{2}} = 12.94\text{ s}$$

(ii) Volumetric capacity Q is calculated using Eq. (4.13)

$$Q = \frac{\begin{array}{c}(1200 \times 2\pi/60\text{ s})^2(0.04\text{ g cm}^{-3})(0.008\text{ cm})^2\\ [(22/7) \times 40\text{ cm}(7.5)^2 - (0.8)^2\text{ cm}^2]\end{array}}{18(0.013\text{ g cm}^{-1}\text{s}^{-1})[\ln(7.5\text{ cm}/0.8\text{ cm})]}$$

$$= \frac{282.84\text{ g cm}^2\text{ s}^{-2}}{0.5237\text{ g cm}^{-1}\text{ s}^{-1}} = 540.08\text{ cm}^3\text{ s}^{-1}$$

(iii) The Σ factor is calculated using Eq. (4.14b)

$$\Sigma = \frac{2(22/7)(40\text{ cm})(7.5\text{ cm})^2(1200 \times 2\pi/60\text{ s})^2}{9.80665 \times 10^2\text{ cm s}^{-2}}$$

$$= \frac{2.235 \times 10^8\text{ cm}^3\text{ s}^{-2}}{9.80665 \times 10^2\text{ cm s}^{-2}} = 22.79\text{ m}^2$$

(iv) Since $t_r = V/Q$, the time required for centrifuging 1000 litres $= (10^6\text{cm}^3)/(540.08\text{ cm}^3\text{ s}^{-1}) = 1851.6\text{ s} = 30.86\text{ min.}$

Questions

1. Explain the principle of centrifugal separation.
2. Write notes on the operation of tubular bowl centrifuge and disc stack bowl centrifuge.
3. What is a gyro tester? What is its use?
4. Explain the significance of RCF value in the selection of centrifuges.
5. Write a note on the scale-up of centrifugation.
6. Discuss the separation of solids by a basket centrifuge.

CHAPTER

5

Adsorption

5.1 ADSORPTION PROCESS

Adsorption is a reversible phenomenon occurring at the surface of a solid. The forces of adsorption are mainly physical and are not strong. Hence desorption of the adsorbate is feasible in physisorption. In contrast chemisorption leads to irreversible adsorption and practically has no application in bioseparations.

The advantages of adsorption process in bioseparations are several. It may be used for primary isolation as well as for concentration of the desired product. Since adsorption is highly selective, the desired product may be adsorbed directly from fermentation broths without the use of any preliminary filtration or centrifugation step. Adsorption does not denature sensitive biomolecules and hence is preferred for isolation of proteins. However, adsorption capacity of any adsorbent is generally small and in addition the design of adsorption process is complicated by non-linear equilibrium and sometimes strong adsorbent-adsorbate interactions.

Solid adsorbents conventionally used in bioseparations include activated carbon of vegetable origin, clay minerals, natural and synthetic zeolites and molecular sieves, alumina, silica gel and ion exchange resins based on synthetic polymers made from styrene and crosslinked with divinylbenzene. Typically, the adsorbents are in the form of small pellets, beads or granules in the range of 0.1–12 mm in size with larger sized particles used in packed beds. The chosen adsorbent must be mechanically stable with relatively a large surface area and chemically inert towards the solute as well as the solvent.

5.2 ADSORPTION ISOTHERMS

Adsorption equilibrium is called an adsorption isotherm and is obtained by plotting solute (adsorbate) concentration in the solid phase as a function of solute concentration in liquid phase at a given temperature. In bioseparations commonly encountered adsorption isotherms belong to any one of the three types, namely, (i) linear, (ii) Langmuir and (iii) Freundlich adsorption isotherms, which are shown in Figure 5.1.

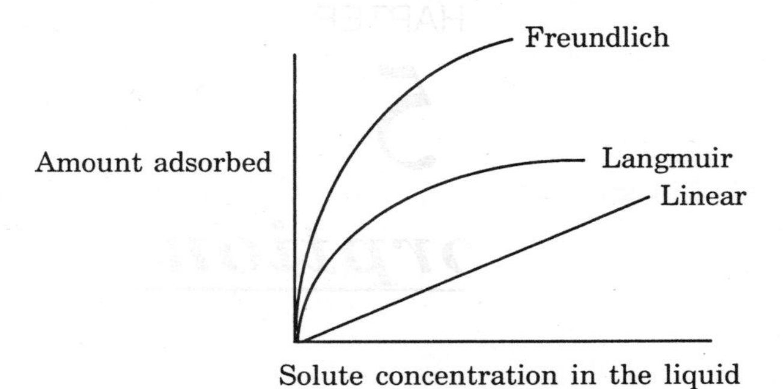

Figure 5.1 Different types of adsorption isotherms.

Linear adsorption isotherm approximates other isotherms over a limited range of solute concentration. It may be quantitatively expressed as:

$$C_a = kS \tag{5.1}$$

where C_a is the amount of solute adsorbed per unit amount of adsorbent (g/g), k is the equilibrium constant and S is the solute concentration in the liquid phase.

Freundlich isotherm is an empirical one but describes the adsorption of a wide variety of antibiotics, steroids and hormones and hence is useful in bioseparations. The isotherms is expressed as:

$$C_a = kS^n \tag{5.2}$$

where n and k are constants determined from experimental data by plotting log C_a versus log S. The slope of the plot gives the dimensionless quantity n, which indicates the favourability of the adsorption process. Adsorption is favourable for $n < 1$ and it is unfavourable if $n > 1$. The dimensions of k depend on the value of n.

Langmuir isotherm has a sound theoretical basis and is often used to describe the adsorption of proteins. The isotherm is expressed as:

$$C_a = \frac{C_0 S}{k + S} \tag{5.3}$$

where C_0 and k are constants determined experimentally by plotting $1/C_a$ against $1/S$. The linear plot gives an intercept equal to $1/C_0$ and a slope equal to k/C_0. The dimensions of C_0 and k are the same as those of C_a and S respectively.

The theoretical basis of Langmuir isotherm involves a reversible reaction between solute and the vacant sites on the adsorbent surface to give filled sites. At equilibrium, the equilibrium constant K is given by

$$\text{Solute} + \text{vacant site} = \text{filled sites} \tag{5.4}$$

$$K = \frac{(\text{Solute}) \times (\text{Vacant site})}{(\text{Filled site})} \tag{5.5}$$

Since the total number of active sites is fixed for a given amount of adsorbent,

$$(\text{Total sites}) = (\text{Vacant sites}) + (\text{Filled sites}) \tag{5.6}$$

combining Eqs. (5.5) and (5.6) we get Eq. (5.7) which is the same as Eq. (5.3).

$$(\text{Filled sites}) = \frac{(\text{Total sites}) \times (\text{Solute})}{K + (\text{Solute})} \tag{5.7}$$

5.3 ADSORPTION TECHNIQUES

Adsorption process can be carried out in batch or continuous mode of operation. We will examine each of these separately.

5.3.1 Batch Adsorption

Batch adsorption is used to adsorb solutes from the liquid phase when the quantities treated are relatively in small amounts. A schematic diagram of the adsorption process is shown in Figure 5.2.

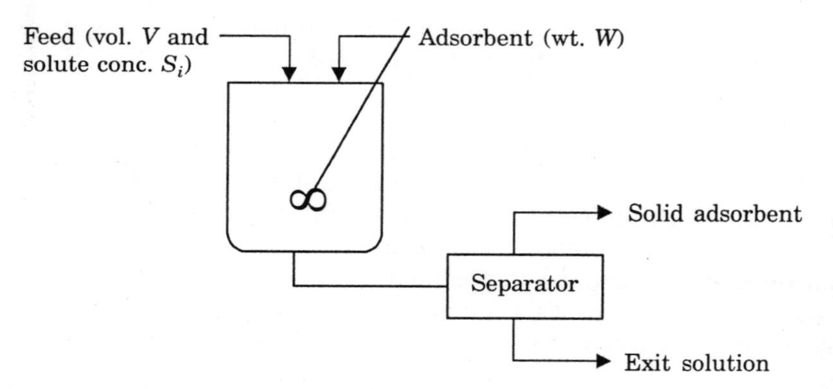

Figure 5.2 Batch adsorption process.

The practice of adsorption for bioseparations involves a series of steps (i) addition of feed solution of the adsorbate to a solid adsorbent contained in a tank. The solute diffuses from the bulk liquid to the gross exterior of surface of the particle (ii) adsorption process which may be selective and slow (iii) removal of spent feed after a time of equilibration and (iv) recovery of the adsorbate from the adsorbent. For analytical treatment of the batch process, an equilibrium relation such as Freundlich or Langmuir isotherm and a mass or material balance are required. In most of the bioseparations, Freundlich isotherm is applicable as given by Eq. (5.2). The mass balance is given by Eq. (5.8).

$$S_iV + C_iW = S_fV + C_aW \tag{5.8}$$

where S_i and S_f are the solute concentration in the feed solution and final concentration at equilibrium respectively, C_i and C_a are the initial and final concentrations of the solute on the adsorbent respectively, V is the volume of the feed solution and W is the amount of the adsorbent. Rearranging Eq. (5.8), we get a linear operating line expressed as Eq. (5.9).

$$C_a = C_i + V/W(S_i - S_f) \tag{5.9}$$

The Eq. (5.2) and Eq. (5.9) may be solved numerically or graphically.

The graphical procedure of plotting C versus S of Eq. (5.9) gives a straight line. The equilibrium isotherm is also plotted on the same graph, to get a curve. The intersection of the two lines gives the final equilibrium values of C_a and S_f as shown in Figure 5.3.

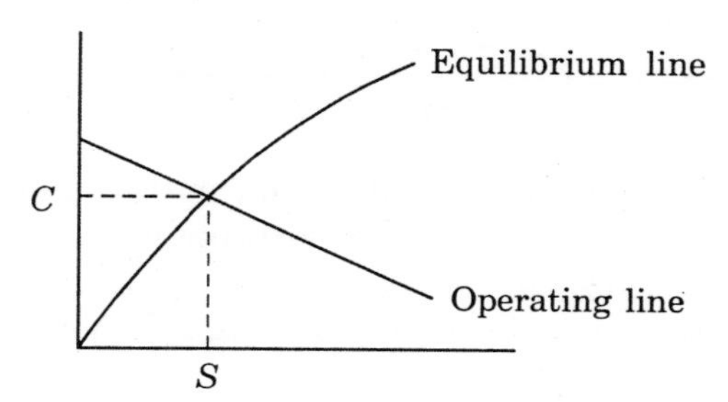

Figure 5.3 Graphical analysis of batch adsorption.

5.3.2 Adsorption in CSTR

Continuous stirred tank reactor (CSTR) is used for large scale separation for fermentation beer processing without resorting to the preliminary step of removal of suspended particulates. The feed enters the tank continuously at a flow rate F and a constant concentration of solute S_i (see Figure 5.4). The tank initially contains pure solvent and fresh adsorbent of known amount W. The concentration of the solute on the adsorbent C varies with time. As the solution flows out steadily at the rate F, the solute exit concentration, S also varies with time initially. However, as the tank is well stirred, the solute concentration in the tank equals that in the outflow.

A qualitative representation of the variation of S with time t is shown in Figure 5.5. Even when there is no adsorption, the solute concentration S in the liquid leaving the tank varies with time. If adsorption is rapid,

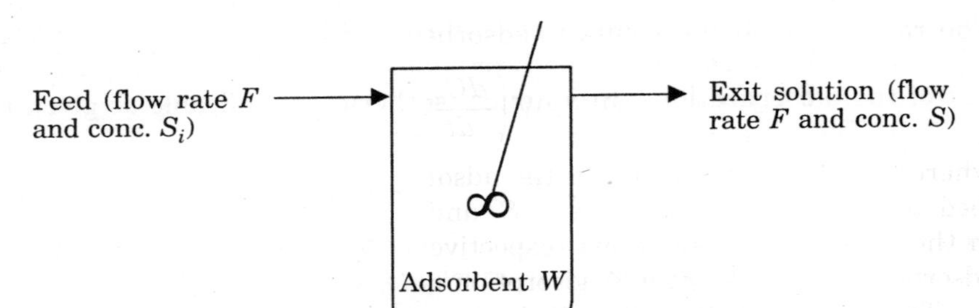

Figure 5.4 Adsorption in CSTR.

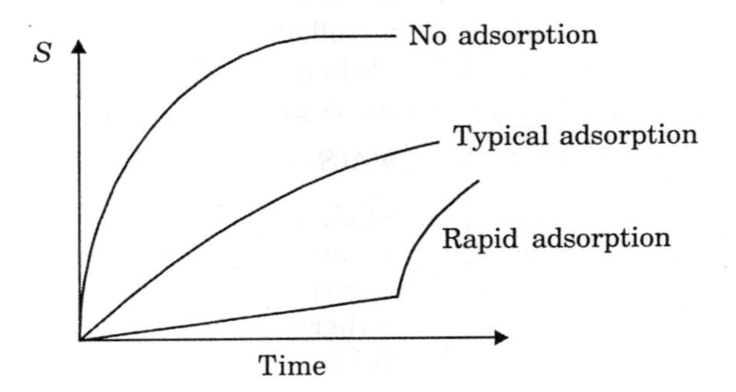

Figure 5.5 Variation of exit solute concentration as a function of time in CSTR adsorption.

Source: P.A. Belter, E.L. Cussler and Wei-Shou Hu, *Bioseparations—Downstream Processing for Biotechnology*, John Wiley & Sons, New York, 1988.

then S will increase slowly with time till the adsorbent is saturated and then increase rapidly for there is not further adsorption. In most cases the rate of adsorption is finite and S varies with time as shown by typical adsorption in Figure 5.5. The quantitative scale-up of CSTR adsorption is rather difficult due to the kinetics of adsorption process, which cannot be inferred from simple equilibrium batch studies. Hence, small-scale studies of CSTR operation is necessary to determine the kinetics of adsorption process for a given adsorbent-adsorbate system.

The analysis of the kinetics in CSTR adsorption takes into account the mass balance of the solute in the liquid and a similar mass balance on the solid adsorbent are given by Eqs. (5.10) and (5.11) respectively. The mass balance of the solute with respect to the liquid can be written as:

(solid entering the tank in the feed) – (solid leaving the tank in the liquid) – (solute adsorbed) = (rate of accumulation of the solid in the tank)

$$\varepsilon V\left(\frac{dS}{dt}\right) = F(S_i - S) - (1 - \varepsilon)V\left(\frac{dC_a}{dt}\right) \tag{5.10}$$

The rate at which the solute is adsorbed is given by Eq. (5.11).

$$(1-\varepsilon)V\left(\frac{dC_a}{dt}\right) = Vr \tag{5.11}$$

where ε is the void fraction in the adsorbent bed, V is the volume of the feed solution, F is the feed rate, S_i and S are the solute concentrations in the feed and exit solutions respectively, C_a is the solute concentration adsorbed and r is the rate of adsorption per volume of the feed (kg m^{-3} s^{-1}).

To find the value of r, it is necessary to know the mechanism responsible for the kinetics of adsorption. In general two limiting mechanisms are commonly proposed: (1) diffusion controlled adsorption that is, adsorption of the solute on the adsorbent is controlled by diffusion of the solute and (2) diffusion controlled adsorption followed by reaction within the adsorbent particles. When adsorption process is diffusion controlled (mechanism 1), the rate is given by Eq. (5.12):

$$r = k'A(S_i - S^*) \tag{5.12}$$

where k' is the mass transfer coefficient and A is the surface area of the adsorbent and S^* is a hypothetical concentration of the solute in solution which would be in equilibrium with the adsorbent; for example, if adsorption follows Freundlich isotherm then S in Eq. (5.2) would be replaced by S^* as given in Eq. (5.13).

$$C_a = k'(S^*)^n \tag{5.13}$$

The mass transfer coefficient k', is a function of stirring in the tank but not a function of temperature.

Equation (5.10) can be rewritten by substituting from Eq. (5.11) as

$$\varepsilon V\left(\frac{dS}{dt}\right) = F(S_i - S) - Vr \tag{5.14}$$

Now, substituting for r from Eq. (5.12) into Eq. (5.14), gives

$$\varepsilon V\left(\frac{dS}{dt}\right) = F(S_i - S) - Vk'A(S_i - S^*) \tag{5.15}$$

Equation (5.11) can be rewritten by substituting from Eq. (5.12) as:

$$(1-\varepsilon)V\left(\frac{dC_a}{dt}\right) = Vk'A(S_i - S^*)$$

$$= Vk'A\left(S_i - \left(\frac{C_a}{k'}\right)^{1/n}\right) \tag{5.16}$$

Equations (5.15) and (5.16) can be solved to get S and C_a.

When diffusion and reaction within the adsorbent particles control the adsorption (mechanism 2), the rate r is given by

$$r = (\sqrt{Dk''}A)(S_i - S^*) \tag{5.17}$$

where D is the diffusion coefficient within the particles and k'' is a reaction constant for adsorption which describes a first order irreversible reaction. The Eq. (5.17) is an approximation but can be used in Eqs. (5.15) and (5.16) to evaluate S and C_a. The expected adsorption rate is independent of stirring in the tank but it is usually a strong function of temperature.

5.3.3 Adsorption in Fixed Bed

A fixed bed reactor is a vertical cylindrical tube filled with adsorbent beads. The fluid containing the solute flows through the packed bed of adsorbent from one end to the other end at a constant flow rate. The situation is more complex than that for a simple stirred tank batch process which reaches equilibrium. Mass transfer resistance is important in the fixed bed process and the process is of unsteady state. The overall dynamics of the system determines the efficiency of the process rather than just the equilibrium considerations. The concentrations of the solute in the fluid phase and on the solid adsorbent change with time and also with the position in the fixed bed as adsorption progresses. Measurements at the end of the column are used to infer the variation of solute concentration with time. At the inlet end of the adsorbent bed the adsorbent is assumed to contain no solute at the start of the process. As the fluid first contacts the inlet end of the bed, most of the mass transfer and adsorption takes place here. As the fluid passes through the bed, the concentration in the fluid decreases very rapidly with distance in the bed to zero even before the outlet end of the adsorbent bed is reached. After a short time, the adsorbent bed at the inlet end is almost saturated and most of the mass transfer and adsorption now takes place at a point slightly away from the inlet end and the mass transfer zone moves farther down the fixed bed. The major part of the adsorption at any time takes place in a relatively narrow mass transfer or adsorption zone. As the fluid continues to flow, this profile moves down further as adsorption progresses. The difference in concentrations is the driving force for mass transfer. The exit concentration of the fluid remains near zero until the mass transfer zone starts to reach the outlet. The exit concentration starts to rise and the saturation of the bed of adsorbent is indicated. The abrupt rise in the solute concentration in the effluent is called the *break point* where the concentration is C_b. The concentration profile in the fluid phase or the variation of solute concentration in the effluent as a function of time obtained by plotting the concentration ratio C/C_0 versus adsorbent bed length (where C is the solute concentration in the liquid at a given time and C_0 is the feed concentration) is called the *breakthrough curve* (see Figure 5.6). After the break point time, the concentration rises rapidly up to point C_d, which is the end of the breakthrough curve where the bed is judged to be saturated. The breakpoint concentration represents the maximum that can be discarded in the effluent and is often taken as

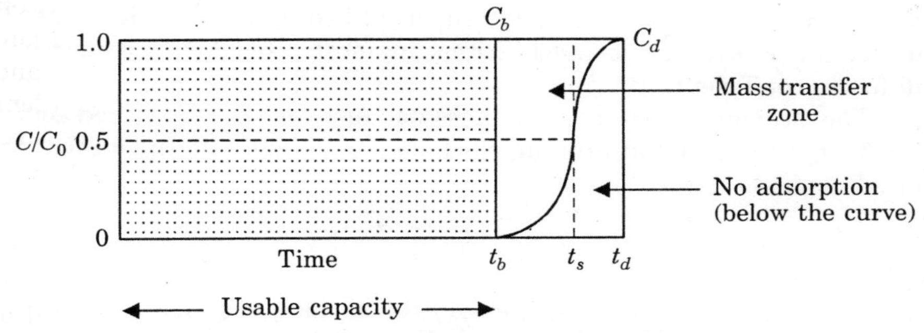

Figure 5.6 Breakthrough curve for fixed bed adsorption.

0.01–0.05 for C_b/C_0. The C_d/C_0 is taken as the point where C_d is approximately equal to C_0. For a narrow mass-transfer zone, the breakthrough curve is very steep and most of the bed capacity is used at the breakpoint. This makes efficient use of the adsorbent and lower energy costs for regeneration. Once the breakpoint is reached, the feed is stopped and the adsorbed material is eluted out using different solvent or pH or ionic strength conditions.

5.3.4 Analysis of Fixed Bed Adsorption Process

The width and shape of the mass transfer zone depend on the adsorption isotherm, flow rate, mass transfer rate to the solid and diffusion in the pores. Though many theoretical methods predict the mass transfer zone and concentration profiles in the bed, the results are approximate and inaccurate and hence laboratory scale experiments are required to scale-up the process.

Analysis of adsorption in fixed bed columns for scale-up is based on unsteady state solute material balance in the fluid and may be described by four basic equations (Eq. (5.18) through Eq. (5.21)). The mass balance on the solute in the liquid is represented by Eq. (5.18).

$$\varepsilon\left(\frac{\partial S}{\partial t}\right) = -v\left(\frac{\partial S}{\partial z}\right) + E\left(\frac{\partial^2 S}{\partial z^2}\right) - (1-\varepsilon)\left(\frac{\partial C_a}{\partial t}\right) \tag{5.18}$$

where ε is the void fraction in the bed, v is the superficial velocity (m/s), and E is axial dispersion coefficient (m^2/s). The left hand side of the equation represents accumulation in the liquid, now taken per differential volume in the bed. The first term on the right hand side of the equation corresponds to the difference in the amounts of solute flowing in and out, the second term represents axial dispersion in the bed and the last term gives the solute transferred from the liquid to the adsorbent. Axial dispersion leads to mixing of solute and solvent even in the absence of adsorption. Diffusion of the solute in fixed bed is not considered as the dispersion coefficient is much larger in most fixed beds compared to diffusion coefficient.

The second equation is a mass balance on the adsorbed solute involving the rate of adsorption, r, which is dependent on the mechanism of adsorption.

$$r = (1 - \varepsilon)\left(\frac{\partial C_a}{\partial t}\right) \tag{5.19}$$

The mechanism may be diffusion controlled mass transfer of the solute from the bulk to the surface of the adsorbent or it may be controlled by diffusion followed by reaction within the adsorbent particles. In either case, the rate of adsorption is assumed to be linear so that the third basic equation for analysis is given as:

$$r = k(S_i - S^*) \tag{5.20}$$

where r represents the rate of adsorption per bed volume and has dimensions of mass adsorbed per volume per time, k is rate constant and S^* is the solute concentration which would exist in the liquid at equilibrium.

The fourth equation of importance gives the equilibrium between the adsorbent and solution concentrations, i.e., an isotherm:

$$C_a = k(S^*)^n \tag{5.21}$$

The solution for the four Eqs. (5.18) through (5.21) is nonlinear and coupled and hence must be determined numerically. However, the numerical solution does not fit experimental results well. Four different strategies have been adopted for a better fit. The first strategy models the breakthrough curves as a ramp. In the second strategy, a characteristic time and a standard deviation are used in modeling the breakthrough curves. In the third strategy, the adsorption equilibrium is assumed to be linear and the equations are solved accordingly while in the fourth strategy, a graphical analysis is used.

5.3.5 Scale-up of Fixed Bed Adsorption

The method to scale-up the fixed bed adsorption column for a given adsorption process is based on the total or stoichiometric capacity of the packed bed tower. When the entire bed comes to equilibrium with the feed, the total capacity of the bed may be taken to be proportional to the area between the curve and the line at $C/C_0 = 1$ in Figure 5.6. The total capacity or the time equivalent to the total capacity or the stoichiometric capacity of the bed, t_t is given by Eq. (5.22).

$$t_t = \int_0^{\infty}\left(1 - \frac{C}{C_0}\right) dt \tag{5.22}$$

The usable capacity of the bed is given by break point time t_b represented by the shaded area in Figure 5.6. The time t_u is the time equivalent to the usable capacity or the time at which the effluent concentration reaches its maximum permissible level as given by Eq. (5.23) and is usually very close to the value of t_b.

$$t_u = \int_0^{t_b} \left(1 - \frac{C}{C_0}\right) dt \tag{5.23}$$

The ratio t_u/t_t is the fraction of the total bed capacity or length utilized up to the break point, H_B (i.e., the length of bed used up to the break point) and is given by Eq. (5.24).

$$H_B = \left(\frac{t_u}{t_b}\right) H_T \tag{5.24}$$

where H_T is the total length of the bed. The length of unused bed, H_{UNB}, is then the unused fraction of total length and may be given as

$$H_{UNB} = \left(1 - \frac{t_u}{t_t}\right) H_T \tag{5.25}$$

H_{UNB} represents the mass transfer section of the bed and depends on the fluid velocity but is independent of total length of the column. The value of H_{UNB} may be calculated at the designed velocity in a small diameter laboratory column packed with the chosen adsorbent by determining the H_B from the breakthrough curve and the total height H_T. Then the full scale adsorbent bed can be designed by first calculating the length of bed necessary to achieve the required usable capacity, H_B, at the break point. The value of H_B is directly proportional to t_b. The total length of full scale adsorbent bed, H_T is obtained by adding the value of H_{UNB} (from lab scale data) to H_B ($H_T = H_{UNB} + H_B$).

The validity of the design procedure depends on the conditions of the laboratory column being similar to those for the full scale unit. The small diameter unit must be well insulated to be similar to the large diameter tower, which operates adiabatically. The mass velocity in both units must be the same and the bed should be of sufficient length to contain a steady state mass transfer zone. Axial dispersion may not be exactly the same in both the lab scale and full scale towers, but if caution is exercised, this method is a useful design method.

An approximate alternative procedure (instead of integrating and obtaining areas) is to assume that the breakthrough curve in Figure 5.6 is symmetrical at $C/C_0 = 0.5$ and t_s. Then the value of t_t in Eq. (5.22) is simply t_s. This assumes that the area below the curve between t_b and t_s is equal to the area above the curve between t_s and t_d.

In the scale-up, it is necessary to consider the change in the column height and the actual throughput of fluid, which might be different from that used in the experimental laboratory unit. Since the mass velocity in the bed must remain constant for scale-up, the diameter of the bed should be adjusted to keep the throughput a constant.

Exercises

1. The lab scale experimental data for the adsorption of an antibiotic on activated carbon are as follows. The solute concentration in the feed (S in mg/cm^{-3}) are 0.3, 0.12, 0.040, 0.018, 0.006 and 0.001 and the corresponding amounts of solute adsorbed on activated carbon (C_a in mg/g) are 0.15. 0.12, 0.095, 0.08, 0.06 and 0.045 respectively. Find out to which adsorption isotherm the data fit.

Solution

The given data does not fit Langmuir adsorption isotherm as the plot of $1/S$ vs $1/C_a$ indicates non-linearity. A plot of log S vs log C_a gives a straight line indicating the applicability of Freundlich isotherm. From the graph the slope (n) is found to be 0.2 and the constant K (intercept) is 0.188. So the equation that fits the experimental data is $C_a = 0.188\ S^{0.2}$.

2. Adsorption of an organic solute on activated silica gel gave the following data after equilibration.

S (mg cm^{-3})	0.139	0.089	0.066	0.047	0.037
C_a (mg/g)	0.03	0.026	0.0225	0.021	0.018

Fit the data to an adsorption isotherm and calculate the constants.

Solution

The given data fits Langmuir isotherm as indicated by the straight line plot for $1/C_a$ versus $1/S$. From the intercept, $C_0 = 0.0385$ and the slope $= n = 0.0246$. The adsorption isotherm is given by $C_a = (0.0385 \times S)/(0.0246 + S)$.

3. The fermentation broth of 1000 litres containing 0.25 g/l of the antibiotic of exercise 1 above, was mixed with 1.5 kg of an adsorbent in a batch reactor and allowed to equilibrate. The values of K and n are 0.188 and 0.2 respectively as shown from the data in exercise 1. Calculate the percent solute adsorbed.

Solution

Equilibrium constraint and mass balance equation are required for analyzing the batch adsorption process by graphical method. The equilibrium constraint is:

$$C_a = 0.188 \times (0.25)^{0.2}$$

Data from exercise 1 with respect to C_a and S are plotted to get the equilibrium line. The mass balance Eq. (5.9) is:

$$C_a = C_i + V/W\,(S_i - S_f)$$

$$= 0 + \frac{1\,\text{m}^3}{1.5\,\text{kg}}\,(0.25\ \text{kg m}^{-3} - S_f)$$

$$= 0.167 - 0.67S_f$$

The data for the mass balance line C_a versus S is obtained numerically and plotted on the same coordinates along with the equilibrium line. From the intersection point of the two lines, C_a and S_f values are obtained as 0.11 and 0.0825 respectively.

The percentage of solute adsorbed $= 100 \times (S_i - S_f)/S_i$

$$= 100(0.25 - 0.0825)/0.25$$

$$= 67\%$$

4. The breakthrough time for $C/C_0 = 0.005$ was determined to be 4.5 hours for a lab scale adsorption column of total length of 20 cm. If a larger scale adsorbent column were to have a breakthrough time of 12 hours what should be the total length of the column for the following lab scale data. The C/C_0 values at breakthrough time and beyond are as follows.

Time (hours)	C/C_0
4.5	0.005
5.0	0.20
5.5	0.50
6.0	0.80
6.5	0.995

Solution

From the graphical presentation of breakthrough curve, the total capacity of the column t_t is calculated from the sum of the areas of usable capacity and mass transfer zone as 4.5 + 1.0 = 5.5 hours.

H_B of the lab scale column $= \left(\frac{t_u}{t_t}\right)H_T = \left(\frac{4.5}{5.5}\right) \times 20 = 16.36$ cm.

H_{UNB} of the lab scale column $= H_T - H_B = (20.00 - 16.36) = 3.64$ cm.

For the larger column the breakthrough time $t_b = 12$ hours.

Hence the new $H_B = (16.36\text{ cm}) \times (12/4.5) = 43.62$ cm

The total length of the larger scale column for achieving the breakthrough time of 12 hours under similar conditions as the lab scale column:

$$H_T = 43.62 + 3.64 = 47.26 \text{ cm}$$

Questions

1. What are adsorption isotherms? How are they obtained?
2. Discuss the principle and practice of batch adsorption.
3. What are the salient features of adsorption in a CSTR? How is the experimental data analyzed?
4. Give an account on the fixed bed adsorption process.
5. Write a note on the scale-up and determination of capacity of a fixed bed adsorption unit.

CHAPTER

6

Extraction

6.1 LIQUID–LIQUID EXTRACTION

Liquid–liquid extraction or solvent extraction as is commonly known, is a classical and versatile method for recovery as well as concentration of a variety of products. The method has been a workhorse in chemical, pharmaceutical and hydrometallurgical industries for over 60 years. However, its use in biotechnology has mostly been for the recovery or isolation and concentration of mainly low molecular weight lipophilic products such as antibiotics and organic acids from fermentation broths. Purification of the desired product also occurs to a certain extent during this unit operation. High molecular weight substances such as antibodies, proteins and enzymes are not amenable to conventional solvent extraction but modified methods such as aqueous two-phase extraction and reverse micellar extraction are highly useful in such cases.

The advantages of solvent extraction include the following.

- Selectivity of extraction directly from fermentation broths or from reaction medium in the case of biotransformations wherein whole cells or enzymes are used for conversion of a substrate into a desired product.
- Reduction in product loss due to hydrolytic or metabolic/microbial degradation as the product is transferred to a second phase with different physical and chemical properties.
- Suitability over a wide range of scales of operation.

However, solvent extraction of biological products is beset with several problems. These include:

(i) Compositional complexity of the fermentation broth due to the presence of a variety of dissolved as well as solid substances, which gives rise to phase complexity and influences the extraction of the desired solute(s).

(ii) The presence of surface active species influences the mass transfer rates.

(iii) The presence of particulate matter and surface active species affects the phase separation.

(iv) Chemical instability of the desired product due to metabolic or microbial activity and also due to the compositional or pH conditions during extraction affects the overall efficiency in the recovery of the desired product.

(v) The rheological properties of the fermentation broths may show time dependence and may be altered affecting the extraction process.

6.2 SOLVENT EXTRACTION PRINCIPLES

Solvent extraction involves the treatment of a large volume of an aqueous solution containing the desired solute with relatively a smaller volume of a non-miscible organic solvent. The solute originally present in the aqueous phase gets partitioned or distributed in both the phases. If the solute has preferential solubility in the organic solvent, more of the solute would be present in the organic phase at equilibrium and extraction is said to be more efficient. The partitioning of the solute between the two phases is expressed quantitatively, on the basis of thermodynamics, by partition coefficient or distribution coefficient. The partition coefficient, K, is the ratio of the solute concentration in the organic phase, C_L, (called the extract phase) to that in the aqueous phase, C_H, (called the raffinate phase), represented by Eq. (6.1).

$$K = \frac{C_L}{C_H} \tag{6.1}$$

The value of K is independent of the solute concentration for a given solvent pair and is a constant at a given temperature. Typical K values determined experimentally for a few biochemicals are given in Table 6.1. The different values of the partition coefficients in a given solvent pair may be conveniently explained on the basis of thermodynamic considerations, particularly, the chemical potential or partial molar free energy, μ of the solute. Extraction is said to attain equilibrium condition when the chemical potentials of the solute in the two phases become equal i.e., $\mu(H) = \mu(L)$. The feed solvent (H) is the denser or heavier aqueous solution and the extracting solvent is usually the lighter organic solvent (L). The chemical potentials may be written as:

$$\mu^0(H) + RT \ln C_H = \mu^0(L) + RT \ln C_L \tag{6.2}$$

TABLE 6.1 Partition Coefficient Data for a few Biochemicals

Solute	*Organic solvent (water is second phase)*	*K* (mol/l) *at 25°C*
Amino acids		
Glycine	*n*-butanol	0.01
Alanine	*n*-butanol	0.02
2-aminobutyric acid	*n*-butanol	0.02
Lysine	*n*-butanol	0.2
Glutamic acid	*n*-butanol	0.07
Antibiotics		
Celesticetin	*n*-butanol	110
Cycloheximide	methylene chloride	23
Erythromycin	amyl acetate	120
Gramicidin	benzene	0.6
	chlorofom-methanol	17
Novobiocin	butyl acetate	100 at pH 7.0
		0.01 at pH 10.5
Penicillin F	amyl acetate	32 at pH 4.0
		0.06 at pH 6.0
Penicillin K	amyl acetate	12 at pH 4.0
		0.1 at pH 6.0

where μ^0 refers to chemical potential in standard reference state, R is gas constant, T is absolute temperature and C_H and C_L refer to the concentrations of the solute in the two phases. Rearranging Eq. (6.2) gives:

$$\frac{C_L}{C_H} = K = \exp \frac{\mu^0(H) - \mu^0(L)}{RT} \tag{6.3}$$

or

$$\ln K = \frac{\mu^0(H) - \mu^0(L)}{RT}$$

Thus the logarithm of the partition coefficient is proportional to the difference in chemical potentials in the standard states. The significance of Eq. (6.3) may be easily explained on the basis of the fact that usually a small amount of the extracting solvent L exists in equilibrium with relatively a large volume of the heavier feed solvent H. At equilibrium, both the phases have dissolved solute. As the amount or volume of the solvent H is in large excess compared to that of L, the chemical potential of the solute in the heavier solvent may be taken as fixed and constant as it does not vary much with a small variation in the solute mole fraction or concentration. The chemical potential of the solute in the lighter phase $\mu(L)$ increases with increasing solute concentration and approaches $\mu^0(L)$

as shown in the Figure 6.1. However, in practice $\mu(L)$ may be limited by the solubility of the solute as indicated by the dotted line. As C_L approaches zero, the value of $\mu(L)$ goes to negative infinity. The intersection of $\mu(L)$ and $\mu(H)$ gives the concentration of the solute in the lighter phase L at equilibrium.

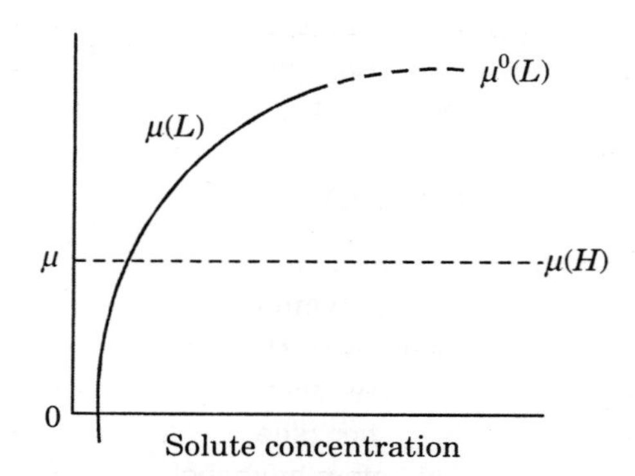

Figure 6.1 Solute chemical potential μ as a function of its concentration.

Source: P.A. Belter, E.L. Cussler and Wei-Shou Hu, *Bioseparations—Downstream Processing for Biotechnology*, John Wiley & Sons, New York, 1988.

It is clear that higher the partition coefficient value for the chosen extracting solvent-aqueous feed system, the greater will be the extraction efficiency. A large density difference between the extractant and the raffinate phase facilitates high throughput particularly when the two phases are to be separated by gravity alone. High viscosity of the solvent on the other hand may affect phase separation. It is necessary that the chosen solvent should have negligible miscibility or solubility in the aqueous feed (or raffinate phase) so that solvent loss will be minimum. It is essential that the solvent is easily recovered and purified for recycling after extraction. The solvent should be easily available and cost effective. A low interfacial tension between the phases facilitates dispersion of the phases and improves mass transfer. However, a low interfacial tension is also disadvantageous in that phase separation is difficult and time-consuming thereby decreasing the throughput.

6.2.1 Selectivity of Extraction

The selectivity of extraction between two solute components is given by separation factor, β, as the ratio of the distribution coefficients of the two solutes A and B.

$$\beta = \frac{K_A}{K_B} \tag{6.4}$$

A higher value of β enables fewer equilibrium stages to achieve a desired degree of purification of A and B.

6.3 EXTRACTION PROCESS

Different extraction processes have been developed over the years. The simplest method is physical extraction. Other methods include dissociative extraction and selective extraction.

6.3.1 Physical Extraction

It involves preferential dissolution of the desired solute in a chosen organic solvent (extracting solvent) for achieving the isolation and concentration. The proper choice of the extracting solvent determines the success of the extraction. The various criteria for selecting the solvent for extraction of a given solute include the partition coefficient, selectivity, density, viscosity, insolubility in the feed solvent, recoverability, cost and interfacial tension of the two phases. Theoretically it is possible to recover the maximum amount of the desired solute by an appropriate choice of the organic solvent. Changing the extracting solvent alters the $\mu^0(L)$ of the solute and thereby enhances the concentration of the solute in the organic phase at equilibrium, i.e. increases the partition coefficient value. However, the major difficulty in physical extraction is to identify an organic solvent, which gives a satisfactorily high partition coefficient value for the desired solute between the extracting and the aqueous phases. Other disadvantages include the high cost, high volatility, flammability and hazardous nature of most of the organic solvents. Some solvents may be even toxic to the desired bioproduct. Thus, the choice of organic solvents satisfying these criteria is highly restricted.

6.3.2 Dissociative Extraction

It overcomes at least partially the difficulty of choosing an organic solvent that satisfies all the criteria. It involves the modification of the physical properties of the solute. Most of the bioproducts such as organic substances and antibiotics are weak acids or bases. The partition coefficient values for the extraction of such solutes into the organic phase is easily enhanced by adjusting the pH of the aqueous phase so that the difference in the dissociation constants of the solute components can be exploited. For example, in the extraction of organic acids or bases the pH of the aqueous feed phase may be adjusted to a value below the pK (negative logarithm of dissociation constant) values of the acids and bases respectively so that the acids or bases exist in undissociated form. The undissociated solute molecules are easily solvated by organic solvents containing ketonic or ester groups through weak hydrophobic interactions. Thus, for a solute which is a weak acid, the partition coefficient K may be expressed as given by the Eq. (6.5).

$$K = \frac{K_i}{(1 + K_a/[H_3O^+]} \quad \text{or} \quad \log_{10}\left(\frac{K_i}{K} - 1\right) = \text{pH} - \text{pK}_a \tag{6.5}$$

where the intrinsic partition coefficient $K_i = [HA]_L/[HA]_H$. The magnitude of K_i is independent of K_a or pH but the value of K is dependent. In strongly acidic solutions $[H_3O^+] >> K_a$ and therefore $K \sim K_i$, which means all of the acid exists as unionized HA. In strongly alkaline solutions $K_a >> [H_3O^+]$ and all of the acid exists as A^- and thus, $K \rightarrow 0$.

Similarly for solutes, which are weak bases the partition coefficient K is given by:

$$\log_{10}\left(\frac{K_i}{K} - 1\right) = \text{pK}_b - \text{pH} \tag{6.6}$$

The changes in partition coefficients due to changes in pH may be used to isolate as well as purify a desired solute. The selectivity of separation, β, between two weak acid solutes is given by

$$\beta = \left(\frac{K_i(\text{A})}{K_i(\text{B})}\right)\left(\frac{1 + K_a(\text{B})/\text{H}^+}{1 + K_a(\text{A})/\text{H}^+}\right) \tag{6.7}$$

Dissociative extraction and the effect of pH on the K value have been used for the extraction of penicillin G from fermentation broths. Penicillin is an extracellular product and exists as dissolved solute. The broth is filtered to remove the biomass and insoluble media components and the pH is adjusted 2.0–2.5 with sulphuric acid to facilitate extraction of penicillin (pK_a = 2.75) into butyl acetate or 4-methylpentanone. Two factors are of importance in this first extraction. These are (i) sensitivity of penicillin to low pH and (ii) tendency to form an emulsion. At low pH penicillin is deactivated, the half-life for deactivation at pH 2.0 being about 0.3 hour. The emulsion formed on the addition of organic solvent is stabilized by the presence of lipids and proteins in the fermentation broth resulting in slow settling and consequently slow phase separation. Hence centrifugal contactors are used for extracting penicillin, which give a short contact time and facilitate rapid phase separation. The product is stripped from the organic solvent by extraction into aqueous buffer of pH 6.0. Acidifying the penicillin-rich aqueous phase and re-extraction into the organic solvent a second time purifies the product. Further purification of the product is achieved by decolourization followed by crystallization.

The separation of penicillin F from penicillin K can be achieved by dissociative extraction in amyl acetate-water system by adjusting the pH as the pK values of the two solutes are 3.51 and 2.77 respectively.

6.3.3 Selective Extraction

Selective extraction of the desired solute, particularly acidic or basic solutes, into the organic phase is achieved by modifying the solute solubility through ion pair or complex or adduct formation.

Ion pair extraction or reactive extraction involves the formation of an organic solvent soluble complex of the desired solute by the use of suitable extractants. Two main types of extractants are commonly used. These include (i) phosphorus bonded oxygen donor extractants such as tributyl phosphate (TBP), tri-octyl phosphine oxide (TOPO), di-2-ethyl hexyl phosphoric acid (DEHPA) used mainly for metal extraction and (ii) amine extractants such as long chain aliphatic amines (e.g. tri-octylamine and di-octylamine) used in the recovery of citric acid from fermentation broths. The extractants are usually dissolved in a diluent (a water immiscible organic solvent). The chosen diluent has to satisfy several criteria such as high selectivity for the desired solute, low toxicity particularly to food and pharmaceutical products, low viscosity and low density to enhance phase separation and stability against degradation (hydrocarbon solvents are relatively more stable than esters or alcohols). The distribution coefficient value of the solute in the chosen diluent-water system should be greater than 1.0 for extraction and less than 0.1 for stripping processes. In addition, the formation of a third phase should be avoided as most of the ion pairs have low solubility in the organic phase and as the ion pair concentration increases a third phase separates out from the organic phase.

Ion pair or reactive extraction has been developed for the recovery of citric acid from fermentation broth. The acid is produced by submerged culture fermentation of molasses by *Aspergillus niger*. The concentration of the citric acid in the broth is about 15% w/v and the use of amine extractant facilitates the extraction of the citric acid into the organic phase. The back extraction of the citric acid from the organic phase into the aqueous phase is carried out at higher temperatures as the K value decreases with increasing temperature.

Penicillin (a weak acid) can be extracted by ion pair or adduct formation at pH 5–6 (penicillin is stable in this pH range with a half-life period of about 92 hours at pH 5) using the extractant N-lauryl-N-trialkyl methylamine dissolved in butyl acetate. Reactive extraction results in reduced loss of penicillin (less than 1%) compared to 20% loss in physical extraction.

The drawbacks of reactive extraction include the potential toxicity of the extractant (particularly the phosphate esters) and the necessity of a back-extraction step to recover the solute in the aqueous phase and recycle the extractant and diluent.

6.3.4 Developments in Extraction Process

A variety of extraction methods have been developed to achieve higher efficiency by altering the mode of contact between the phases and transfer of the solute to organic phase and re-extraction (or stripping) into the aqueous phase. The methods include (i) liquid emulsion membrane extraction (ii) supported liquid membrane extraction and (iii) predispersed solvent extraction.

Liquid emulsion membrane extraction involves the transfer of the desired solute from an outer aqueous phase through a solvent phase into an inner aqueous phase and thereby combining the two extraction processes (extraction and stripping) into a single step. The bulk aqueous feed phase is separated from the dispersed aqueous phase by an organic solvent (membrane phase). The inner aqueous phase is emulsified, with an added emulsifier, in the membrane phase under high shear forming droplets of 1–5 μm in diameter. The transfer of the desired solute from the outer feed phase through the membrane into the inner aqueous phase is due to purely physical mechanism or may be facilitated by the use of extractants in the membrane. This method has been successfully tested for citric acid extraction into an inner aqueous phase of sodium carbonate and an emulsifier (Span 80) separated by the membrane phase of tri-octylamine (extractant) in n-heptane (diluent).

Supported liquid membrane extraction involves the transfer of the solute from an aqueous phase across the organic solvent phase supported on a hydrophobic support to a second aqueous phase. Advantages of this method include low solvent requirement, absence of emulsification problems and potential to use more selective or expensive extractants.

Predispersed solvent extraction involves the use of micron or sub-micron sized droplets, called colloidal aphrons, of an organic solvent stabilized by a surfactant surface layer, dispersed in the continuous aqueous phase. The concentration of surfactant is used to control the size of the aphrons. Addition of a suspension of aphrons to water disperses into individual aphrons which provide a large surface area for mass transfer during extraction. Slow settling rates and difficulty in filtration to recover colloidal aphrons are the disadvantages.

6.4 EQUIPMENT FOR EXTRACTION

The type of equipment used decides the success of the extraction. In general, a high degree of turbulence facilitates intimate contact between the two liquid phases and allows a high rate of mass transfer. After sufficient time of contact the two phases must be separated quickly. The density difference between the two phases is not large and hence the selection of equipment becomes more critical for the success of extraction. Two main types of equipment find use in solvent extraction. These include (i) vessels in which mechanical agitation facilitates mixing and (ii) vessels in which mixing is done by the counter-current flow of the two liquids themselves. Biological extractions are affected adversely by several factors such as high viscosity of the feed, low density difference between the aqueous feed and the organic solvent, high solids content and the presence of surface active species which facilitate emulsification. In addition, extraction from fermentation broths require prior separation of the suspended solids by filtration or centrifugation and disposal of the filter cake. During such solid–liquid separation, considerable amount of the desired solute is also lost as part of the filtrate held in the cake. Centrifugal contactors, in which the counter-current flow of the two phases facilitates mixing, have been widely used in antibiotic extraction to overcome these problems. Centrifugal contactors simultaneously disperse and separate the phases in a short time and hence minimize the degradation of labile products. Two important factors that control the operation of the centrifugal contactors include the interface position and increase droplet sedimentation velocity in a centrifugal field. The interface position decides which phase is dispersed in the contactor. The interface position is related to the outlet positions of the light and heavy phases, which in turn depend on the hydrostatic balance within the contactor. The centrifugal field influences the coalescence and the mass transfer characteristics by increasing the droplet sedimentation velocity.

Two main types of centrifugal contactors find use in biological extractions. These include (i) single stage disc stack separators and (ii) multi-stage differential contactors.

In the disc stack separator, the mixture of the feed broth and the solvent is pumped to the disc stack contained in a bowl centrifuge. Phase separation is achieved by a collection of discs stacked with a narrow spacing between them. The configuration allows the separation of three phases, the lighter and heavier liquid phases and solids. The heavier phase is forced to the rim of the bowl while the lighter liquid phase moves towards the rotation axis of the bowl. Heavier solids are also thrown to the rim of the bowl and are removed through discharge nozzles. Interfacial solids accumulate at the interface of the liquids at the center of the disc stack and ultimately clog the disc channels.

The differential contactor consists of a rotor containing a series of concentric perforated plates or a long spiral channel fitted with baffles and screens to facilitate mixing of the phases. The lighter phase is fed at the rim of the rotor while the heavier feed is introduced at the centre close to the rotor axis. As the rotor rotates the heavy liquid is forced down through the holes in the plates or via the spiral channel resulting in the dispersion of the two phases. This counter-current movement brings about a multi stage extraction. The product stream is collected along the rotation axis. The differential contactor lacks a discharge system for solids and hence the successful operation depends on the solids remaining in suspension in the aqueous phase.

The decantor extractor originally developed for dewatering of solids modified for liquid–liquid extraction is claimed to be advantageous as it can tolerate higher solids content in the feed. The centrifuge consists of a bowl with separate extraction and clarifying zones. The solvent and the aqueous feed are fed at the opposite ends of the extraction zone of the bowl centrifuge along the rotation axis. The two phases move counter-current to each other within the bowl. The heavy phase along with the solids is thrown to the rim of the bowl and collected from one end of the bowl. The lighter phase remains close to the rotation axis and is pumped out from the opposite end of the bowl.

6.5 OPERATING MODES OF EXTRACTION

Two major modes of extraction practiced commonly in the industry include the batch extractions and continuous extractions. Batch extraction may be carried out in single stage or in multiple stage. Continuous extraction includes two types, co-current and counter-current modes of extraction. The different modes of extraction are discussed in this section.

Apart from the thermodynamic aspects (partition coefficient), kinetics need to be considered in large scale extractions.

6.5.1 Batch Extraction

In a single stage batch extraction, the aqueous feed is mixed with the organic solvent and after equilibration, the extract phase containing the desired solute is separated out for further processing. The magnitude of the separation factor should be large for a successful single stage batch extraction. A schematic representation of the single stage batch operation is shown in Figure 6.2.

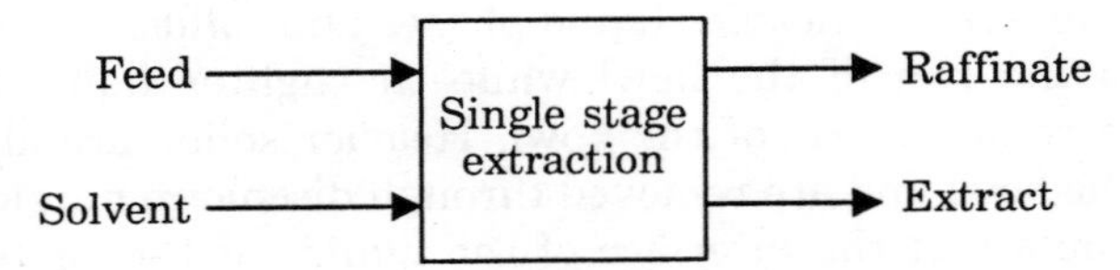

Figure 6.2 Schematic representation of a single stage batch extraction.

The equilibrium concentrations of the solute in the two phases may be calculated by either of the two approaches, namely, (i) analytical method or (ii) graphical method in order to determine the efficiency of extraction.

Analytical method. If the initial amount of the solute, x, in the feed aqueous phase is assumed to be C_0 moles and after extraction C_1 moles, then the concentrations of the solute in the extract phase, L, and the feed phase H, at equilibrium are given by Eqs. (6.8) and (6.9) respectively.

$$C_L = \frac{C_0 - C_1}{V_L} \tag{6.8}$$

$$C_H = \frac{C_1}{V_H} \tag{6.9}$$

where V_L and V_H are the volumes of the two phases. The partition coefficient K may be given as:

$$K = \left[\frac{(C_0 - C_1)/V_L}{C_1/V_H}\right] \quad \text{or} \quad K\left(\frac{V_L}{V_H}\right) = \frac{C_0 - C_1}{C_1} \tag{6.10}$$

The concentrations of the solute in the two phases at equilibrium may be obtained as:

$$C_1 = \frac{C_0}{1+E} \quad \text{and} \quad (C_0 - C_1) = \frac{KC_0}{1+E} \tag{6.11}$$

where E is the extraction factor given by,

$$E = K\left(\frac{V_L}{V_H}\right) \tag{6.12}$$

If K is large, most of the solute will be extracted into the extraction solvent and the magnitude of E will be large.

Graphical method. The graphical method to arrive at equilibrium concentrations of the solute in batch extraction is useful for engineering analysis, particularly when the equilibrium is complex. The method depends on the two basic relations of partition coefficient (equilibrium constraint) and mass balance. The equilibrium constraint may be non-linear and a complex function as given in Eq. (6.13).

$$x = x(y) \tag{6.13}$$

and the mass balance relation as given by:

$$V_H y_i + V_L x_i = V_H y_f + V_L x_f \tag{6.14a}$$

where x_i (= 0) and x_f are the initial and final concentrations of the solute in the lighter phase and y_i and y_f are the initial and final concentrations of the solute in the feed (heavier) solvent. Since before extraction $x_i = 0$ the above equation may be simplified as:

$$x_f = \left(\frac{V_H}{V_L}\right)(y_i - y_f) \tag{6.14b}$$

The Eqs. (6.13) and (6.14b) are plotted on the same coordinates (see Figure 6.3). The intersection of the equilibrium line (energy balance) and the mass balance line (operating line) gives the values of x and y after extraction.

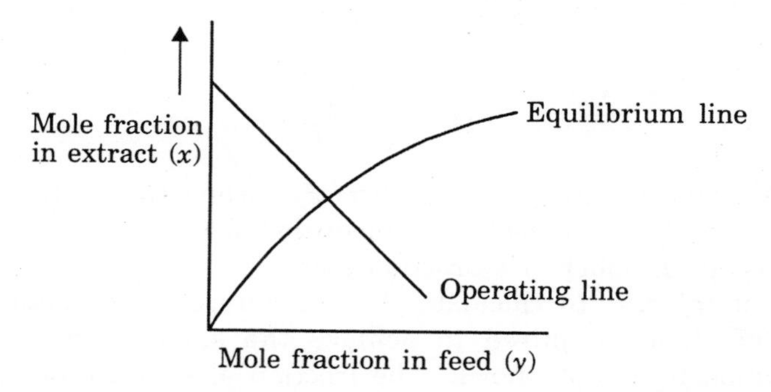

Figure 6.3 Graphical method of analysis for a batch extraction.

6.5.2 Batch Extraction in Multiple Stage

A single extraction may not be sufficient to remove the solute quantitatively from the aqueous phase into the organic phase. Hence a fresh volume of organic solvent is contacted with the raffinate (aqueous phase after first extraction) and allowed to equilibrate. This second extraction may be followed by a third extraction and so on giving rise to batch extraction in multiple stage. The batch extraction in multiple stage is simply the repetition of single stage batch extraction. The schematic diagram of the staged extraction is given below. Figure 6.4. shows the schematic representation of the cross-current or multi-stage batch extraction.

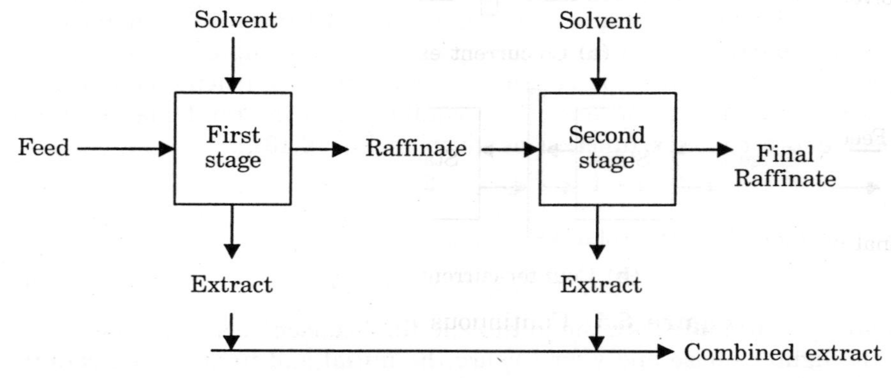

Figure 6.4 Schematic representation of a two-stage batch extraction.

The solute concentration in the aqueous phase after second extraction C_2 is given in terms of solute concentration C_1 (final concentration of first extraction and initial concentration of second extraction) or in terms of C_0 as

$$C_2 = \frac{C_1}{1+E} \quad \text{or} \quad C_2 = \left(\frac{C_0}{1+E}\right)^2 \tag{6.15}$$

In general for multiple stage extraction of n stages, the general equation is:

$$C_n = \left(\frac{C_0}{1+E}\right)^n \tag{6.16}$$

The final concentration C_n is directly proportional to the initial concentration C_0 multiplied by a separation factor raised to nth power, where n is the number of extraction stages.

It is of interest to calculate the percentage of extraction and the number of stages required to achieve the same even if the initial concentration C_0 is not known. The fraction of solute remaining in the aqueous phase after extraction is given as mole fraction.

$$f = \frac{C_n}{C_0} \text{ or as percentage } f = 100\left(\frac{C_n}{C_0}\right) \tag{6.17}$$

6.5.3 Continuous Extraction

Continuous extraction can be co-current or counter-current modes as shown in Figure 6.5.

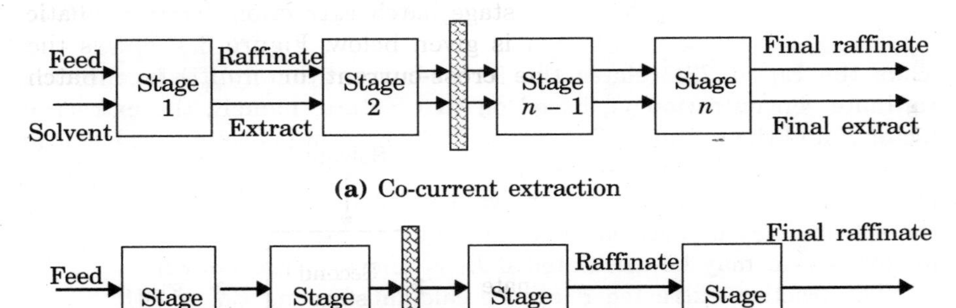

(a) Co-current extraction

Feed
Stage n
Stage $n - 1$
Stage 2
Raffinate
Extract
Stage 1
Final raffinate
Solvent
Final extract

(b) Counter-current extraction

Figure 6.5 Continuous extraction modes.

Counter-current mode of extraction is the most widely used mode of operation. Each stage in the cascade is identified by a number, starting from stage 1 on the right side to stage n on the extreme left. The feed enters on the left side and the pure solvent on the right of this end feed mode of operation. The concentrations of the solute are specified by the stage from which it is leaving; for example, the solvent leaving the first stage has a solute concentration x_1 and the raffinate leaving the second stage and flowing into the first stage has a solute concentration y_2. Thus the feed entering at the left side has a solute concentration of y_{n+1} as the initial feed concentration.

The analytical method of evaluating the performance of the cascade is based on considering the equilibrium and mass balance constraints. Equilibrium is assumed to be achieved at each of the n stages as represented by Eq. (6.18). The mass balance for the first stage is given by Eq. (6.19).

$$x_n = Ky_n \tag{6.18}$$

$$V_H y_2 + V_L x_0 = V_H y_1 + V_L x_1 \tag{6.19}$$

Combining the above two equations we get:

$$y_2 = (E + 1)y_1 \tag{6.20}$$

where $E = K(V_L/V_H)$. For the second stage, the mass balance is:

$$V_H y_3 + V_L x_1 = V_H y_2 + V_L x_2 \tag{6.21}$$

and therefore

$$\begin{aligned} y_3 &= (1 + E)\, y_2 - E y_1 \\ &= (1 + E + E^2)\, y_1 \end{aligned} \tag{6.22}$$

For n stages

$$y_{n+1} = (1 + E + E^2 + \ldots + E^n) y_1 \quad \text{or} \quad y_{n+1} = y_1 \frac{E^{n+1} - 1}{E - 1} \tag{6.23}$$

Thus the Eq. (6.23) relates the feed concentration (y_{n+1}) to the final raffinate concentration y_1, based on two factors, namely, the extraction factor E and the number of stages n. The equation is useful in calculating the exit concentration y_1 if the values of n, E and y_{n+1} are known. If the ratio y_1/y_{n+1} and n are known, the value of E and the flow of the feed and the solvent may be arrived. Alternatively, the number of stages required for extraction may be calculated if E, y_1 and y_{n+1} are known.

The fraction extracted F can be calculated using Eq. (6.24).

$$F = \frac{V_L x_n}{V_H y_{n+1}}$$

or

$$F = 1 - \frac{y_1}{y_{n+1}}$$

or

$$F = \frac{E(E^n - 1)}{E^{n+1} - 1} \tag{6.24}$$

As E becomes large, F approaches 1 or if $E = 1$ then $F = n/(n+1)$ as $E \rightarrow 0$ and $F \rightarrow 0$.

In the graphical method, two basic relations required include the statement on equilibrium and the mass balance given by Eqs. (6.25) and (6.26) respectively.

$$x_n = x_n(y_n) \tag{6.25}$$

$$x_n = \frac{V_H}{V_L}(y_{n+1} - y_1) \tag{6.26}$$

Plotting these equations on the same coordinates gives the graphical representation as shown in Figure 6.6.

The point (y_1, $x = 0$) on the graph corresponds to the right end of the cascade of the counter-current extraction. The point (y_1, x_1) on the equilibrium line corresponds to the concentrations of the solute in the first stage while the point (y_2, x_1) on the operating line is equivalent to the mass balance in the first stage. Repeating the reading procedure vertically on the equilibrium line and horizontally on the operating line till a point (y_{n+1}, x_n) equals or exceeds the feed concentration gives the number of stages required for the extraction.

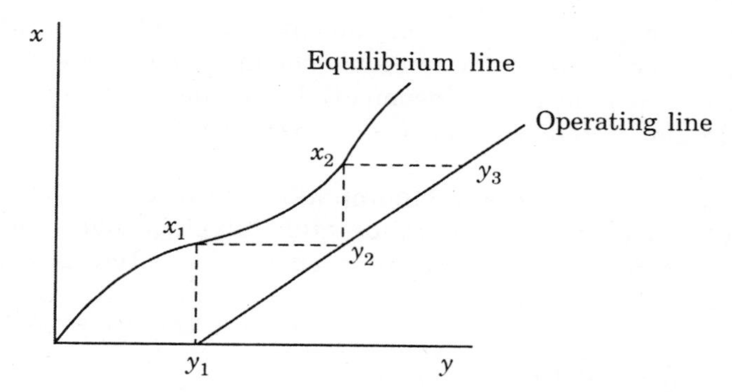

Figure 6.6 Graphical analysis of counter-current extraction.
Source: P.A. Belter, E.L. Cussler and Wei-Shou Hu, *Bioseparations—Downstream Processing for Biotechnology*, John Wiley & Sons, New York, 1988.

6.6 AQUEOUS TWO-PHASE EXTRACTION

To extract biologically active polymers such as proteins, enzymes and nucleic acids it is necessary to have a two-phase system, which will not denature the biopolymers. It is in this context that aqueous two-phase extraction is of immense importance in biotechnology.

In aqueous bi-phase extraction, the basic principle involves the differential partitioning of desired protein(s) in two immiscible phases extracted from an aqueous feed containing a complex mixture of proteins. Both the phases contain distinct polymers or polymer and salt and are relatively rich in water (80–85% content) and do not denature proteins and other biopolymers.

6.6.1 Aqueous Two-phase (or Bi-phase) Systems

Aqueous solutions of most hydrophilic polymers such as starch, gelatin or agar show incompatibility in binary mixtures and form two immiscible aqueous phases, each containing primarily only one of the two phase-forming polymers and a high proportion of water. At low concentrations of the polymers, homogeneous solutions (single phase) are obtained, but at discrete concentration ratios, measurable by cloud point method, phase separation occurs. The mutual insolubility of the two hydrophilic polymers may be attributed to the molecular form of each polymer, which results in mutual repulsion. Polyethylene glycol is a linear chain polymer with a high density of lone pairs of electrons. In contrast, dextran is globular and does not have any tendency for dipole formation. Each polymer tends to attract molecules of similar shape, size and polarity and repel molecules of different type forming separate phases.

A similar phase separation occurs with a mixture of aqueous solutions of a hydrophilic polymer and a low molecular weight substance such as potassium phosphate and sodium chloride. A mixture of aqueous solutions of polyelectrolyte and a non-ionic polymer also shows phase separation. A list of such mutually incompatible polymer–polymer and polymer-low molecular weight substance systems are listed in the Table 6.2.

TABLE 6.2 Mutually Incompatible Polymer–Polymer, Polymer-low Molecular Weight Substance and Polyelectrolyte-non-ionic Polymer + Salt Systems Forming an Aqueous Two-phase System

Polymer	*Polymer/Low molecular weight substance*
1. Polymer–Polymer systems	
Polyethylene glycol (PEG)	Dextran
	Polyvinyl alcohol (PVA)
	Polyvinyl pyrrolidone (PVP)
	Ficoll
Polypropylene glycol	Dextran
	Polyvinyl alcohol
	Polyvinyl pyrrolidone
Polyvinyl alcohol	Dextran
	Methyl cellulose
Polyvinyl pyrrolidone	Dextran
	Methyl cellulose
2. Polymer-low molecular weight substance systems	
Polyethylene glycol	Potassium phosphate
Polyvinyl pyrrolidone	Potassium phosphate
Polypropylene glycol	Glucose
Dextran	Propyl alcohol
Sodium dextran sulphate	Sodium chloride
3. Polyelectrolyte-Non-ionic Polymer + salt systems	
Sodium dextran sulphate	Polyethylene glycol and sodium chloride
	Polyvinyl alcohol and sodium chloride
Sodium carboxy methylcellulose	Polyethylene glycol
	Polyvinyl alcohol

6.6.2 Phase Diagrams

The miscibility range and the formation of aqueous two-phase systems between two phase forming polymer–polymer or polymer-solute combinations depends on their concentration ratios and may conveniently be discussed with the help of phase diagrams. A phase diagram of a typical two-phase aqueous polymer system of polyethylene glycol and dextran is shown in Figure 6.7. Both the polymers are separately miscible with water in all proportions and with each other at low concentrations. However, as

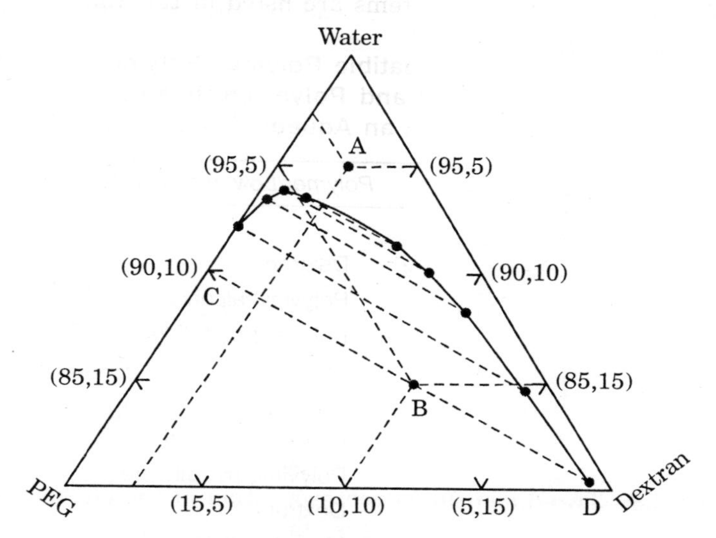

Figure 6.7 Phase diagram of an aqueous two-phase system.

the polymer concentrations increase, phase separation occurs between the PEG-rich upper phase and dextran-rich lower phase, each phase consisting of more than 80% water. The curve separating the two regions of bi-phase and the homogeneous phase is called the binodal curve. Within the bi-phase region, any mixture of the three components, PEG-dextran-water, splits into two distinct phases as governed by the intersections of the tie-line passing through the mixture point with the binodal.

For example, a mixture of the three components of a certain composition, say, 95% water + 2.5% PEG + 2.5% dextran, represented by point A in the phase diagram (Figure 6.7) lies above the binodal curve and hence exists as a homogeneous phase. On the other hand, a mixture represented by point B with a composition of 85% water + 5% PEG + 10% dextran, is below the binodal curve and is not stable as a homogeneous phase. It will separate into two phases, one phase with 90% water + 10% PEG (represented by point C in the phase diagram) and a second phase with 80% water + 19% dextran + 1% PEG (represented by point D in the phase diagram). The relative volumes of the two phases are given (by inverse lever arm rule) as V_C/V_D = CB/BD, where V_C and V_D are the volumes of the phases with compositions represented by points C and D respectively and CB and CD are the lengths of the tie-line segments connecting point C with B and point B with D respectively. All mixtures represented by points on the same tie-line give rise to identical phase systems but with different volume ratios.

Since water is the major component in both the phases, the polymer concentrations may be expressed in terms of weight percentage of the polymer in water. A two-dimensional phase diagram can be easily constructed (see Figure 6.8) instead of the conventional triangular presentation of the ternary system.

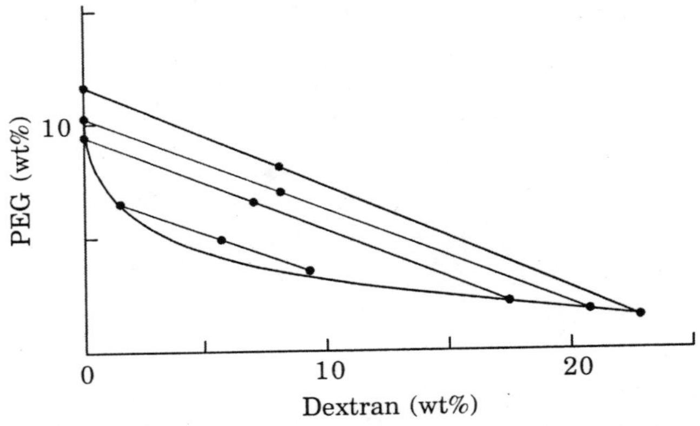

Figure 6.8 Two-dimensional representation of the aqueous two-phase system.

The properties of aqueous bi-phasic systems depend on a variety of factors such as molecular weights of the polymers, viscosity, density, interfacial tension and temperature. Molecular weights of the polymers involved determine the concentrations of the polymers necessary to split a homogeneous solution into two phases. High molecular weight polymers give biphasic system at relatively low concentrations with good phase separation. If the difference in the molecular weights of the two polymers is large, the binodal solubility curve is more asymmetrical.

The viscosities of the two phases are important in bringing about contact between the phases and mass transfer during extraction and in phase separation. The presence of various constituents in the fermentation broth contribute to the viscoelastic properties of the broth and can influence the kinetics of contact and mass transfer during extraction of desired proteins directly from fermentation broths.

Difference in density between the two phases is preferable for better contact and phase separation. However, the density ratio of the two phases is mostly close to unity in most cases. Phase inversion, i.e., change of the heavier phase into lighter one is unusual.

A low interfacial tension is preferable for effective dispersion and creation of high interfacial area. In most two-phase systems, the interfacial tension is low in the range of 10^{-2}–10^{-3} mN m^{-1} for PEG-dextran systems and 10^{-1}–10^{-2} mN m^{-1} for PEG-salt mixtures.

Temperature influences the contacting process through changes in viscosity and phase equilibrium. Thus, increase in temperature decreases the miscibility and enhances phase separation in PEG-potassium sulphate-water system. In contrast, the miscibility of PEG-dextran system increases with increase in temperature.

6.7 THEORETICAL PRINCIPLES OF AQUEOUS TWO-PHASE EXTRACTIONS

The principles of extraction in aqueous two-phase systems may be explained in terms of partition of the solute between the two phases. The distribution of biopolymers between two water rich liquid phases may be attributed to charge interaction, hydrogen bonding and van der Waals interactions between solute molecules and the polymer molecules of the liquid phases. These interactions are influenced by various parameters such as molecular weights of the polymers, type and concentration of the salt, pH and temperature. The partitioning coefficient, K, may be expressed as the ratio of the concentration of the desired protein in the lighter upper phase to that in the denser lower phase.

The partition coefficient may be expressed as a function of five important factors involving electrical (K_{elec}), hydrophobic (K_{hphob}), hydrophilic (K_{hphil}), conformational (K_{conf}) and ligand (K_{lig}) type interactions as given by:

$$\ln K = \ln K_{elec} + \ln K_{hphob} + \ln K_{hphil} + \ln K_{conf} + \ln K_{lig} \quad (6.27)$$

Typical values for partition coefficients for cells, cell fragments and DNA are in the range of 100 while for most proteins and enzymes the values are in the range of 10.

The partition coefficient for a protein in aqueous two-phase extraction is influenced by factors such as the molecular weight of the phase-forming polymer, type of salt and its concentration and pH.

The effect of polymer molecular weight on the partition coefficient is nicely demonstrated in the selective extraction of fumarase from a protein mixture in PEG-potassium phosphate system containing varying proportions of PEG 4000 and PEG 400 in the aqueous polymer phase. Fumarase is preferentially extracted into the aqueous PEG phase, the K values increasing with increasing PEG 400 content in the mixture PEG 4000 + PEG 400. The partition coefficient values for the contaminant proteins in the feed broth also show an increasing trend particularly at PEG 400 above 12%. Selective extraction of fumarase is possible by adjusting the molecular weight profile of PEG by simply mixing the requisite amounts of the polymer of known molecular weight. Thus, increasing the polymer molecular weight in one phase decreases the partitioning of the protein to that phase. A similar effect was observed with a decrease in partitioning of pullulanase in PEG phase of PEG-dextran system. The K values decreased from 1.3 to 0.25 as the molecular weight of PEG increased from 1500 to 6000. At higher PEG molecular weights, the K values were not affected. In general, the dependence of protein partitioning is small for small proteins of about 10,000 Da but increases with large proteins of molecular weights in the range 250,000 Da. This effect may be due to the altered phase compositions.

The type of salt used for phase formation and its concentration as well as the presence of an affinity ligand (e.g. Cibacron blue dye) in one of the phases influences the partition coefficient values as shown in the case of glucose-6-dehydrogenase extraction in PEG-Cibacron blue-salt system. The effect of pH on the K values for L-hydroxyisocaproate dehydrogenase partitioning in PEG-potassium sulphate system showed that the K values vary between a low value of less than one at pH 5 to a maximum of about 3 at pH 7.2.

6.8 AQUEOUS TWO-PHASE EXTRACTION PROCESS

The extraction process involves two steps (i) mixing the aqueous feed with the two-phase solvent system for extraction and (ii) phase separation. The recovery of the desired product is based on a sequential stepwise removal of products and impurities by adjusting the physical and chemical compositional conditions of the two phases and may be explained with a flow chart (see Figure 6.9). The first step in the separation of a desired protein (or an enzyme) directly from a fermentation broth is the removal

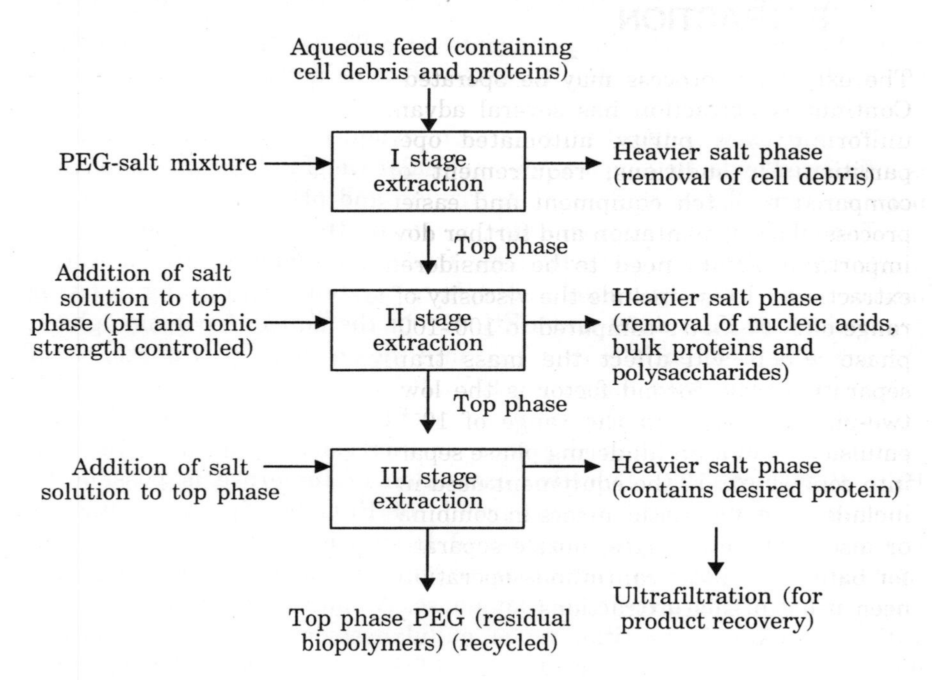

Figure 6.9 Flow chart for an aqueous two-phase extraction.

of cell debris into the heavier phase e.g., the salt phase in PEG-potassium phosphate biphasic system by contacting the aqueous feed with the solvent system. The proteins and other biopolymers are extracted into the PEG rich lighter phase. After centrifugal separation of the two phases, the lighter phase is contacted with a solution of the salt at fixed pH and fixed ionic strength. This second stage of extraction removes nucleic acids, polysaccharides and bulk proteins into the heavier salt phase leaving the almost pure protein or enzyme in the lighter PEG-rich phase. The two phases are separated by centrifugation. Further purification of the desired protein or enzyme may be achieved by a third stage extraction of the protein or enzyme into the salt phase by adding salt solution and manipulating the pH and ionic strength conditions. The desired protein or enzyme may be recovered from the heavier salt phase by ultrafiltration. The protein extracted into the polymer rich phase can also be recovered by chromatography. PEG as well as the salt phases may be recycled. Recovery of solvent phase, particularly with polymers such as PEG and dextran, is necessary for both commercial (cost of the polymers) and environmental reasons. The polymer may be recovered by simply transferring the protein into salt-rich phase by adjustment of pH or ionic strength as the last step of aqueous two-phase extraction.

6.9 EQUIPMENT FOR AQUEOUS TWO-PHASE EXTRACTION

The extraction process may be operated in batch or continuous mode. Continuous extraction has several advantages such as greater product uniformity and purity, automated operation and better control of partitioning conditions, requirement of smaller process equipment compared to batch equipment and easier integration of the extraction process with fermentation and further downstream processing steps. Two important factors need to be considered in large aqueous two-phase extractions. These include the viscosity of the extract phase being in the range of 2–10 mPa s compared to 100–10000 mPa s for broth homogenate phase which will affect the mass transfer kinetics and also phase separation. The second factor is the low interfacial tension in aqueous two-phase systems in the range of 10^{-4} dyne/cm, which will facilitate emulsion formation hindering phase separation. Taking both these factors into consideration, the equipment used in aqueous two-phase extractions include in-line or static mixers in combination with centrifugal separators or disc-stack centrifuges, nozzle separators or decantor extractors, used for batch as well as continuous operations. Column contactors have also been used in such extractions.

6.10 APPLICATIONS OF AQUEOUS TWO-PHASE EXTRACTION

Recovery of enzymes. Recovery of fumarase, penicillin acylase and formate dehydrogenase from fermentation broths are some of the examples of the applications of aqueous two-phase extractions.

Fumarase was recovered from baker's yeast homogenate and also from the *B. ammoniagenes* fermentation broth in a two-stage extraction process. After fermentation, the cell suspension was homogenized in a bead mill to release the intracellular fumarase. In the first stage of downstream processing, the supernatant was mixed with PEG and potassium phosphate mixture to create the aqueous two-phase system in disc stack separators. The enzyme in the supernatant was extracted into the PEG rich aqueous phase (top phase) while the bottom phase containing cell debris and proteins was discarded after separation. In the second stage, the enzyme was re-extracted from the PEG rich phase into the salt phase (bottom phase) by adding more salt so that the PEG could be recovered for recycling.

Penicillin acylase was isolated by a similar two-stage process from the fermentation broth of the recombinant *E. coli* strain.

Formate dehydrogenase (FDH) of about ten-fold concentration and about 70% purity was achieved in a four-stage process involving aqueous two-phase extraction. The first stage involved the preparation of the cell homogenate after completion of the fermentation process. The biomass was harvested and disrupted by bead mill to release the FDH and the cell homogenate was selectively denatured by heating to enhance the specific enzyme activity by about 60%. The biomass suspension was contacted with PEG/potassium phosphate/water extractant and the phase separation was achieved in a continuous nozzle separator. The protein rich supernatant (top PEG phase) had a high degree of purity of protein close to 100%. The cell debris contained in the heavy raffinate phase was discarded. In the second stage, the protein rich supernatant was contacted with a mixed electrolyte solution of potassium chloride and potassium phosphate to partition the nucleic acids in the heavy raffinate phase separated by gravitational settling leaving the FDH in the supernatant. In the third stage, the supernatant was contacted with a fresh extractant of PEG 400/PEG 1500/potassium phosphate. Careful adjustment of salt concentration and the molecular weight ratios of the PEG selectively partitioned the FDH into the heavier phase. The bulk proteins were partitioned into the lighter phase for rejection. The fourth and final stage involved the polishing operation to purify FDH by removing any residual impurities. This was achieved by adding potassium phosphate solution to the FDH stream. The FDH was re-extracted into the lighter PEG phase while the impurities of residual nucleic acids, proteins and polysaccharides, were extracted into the salt-rich heavier raffinate phase.

Recovery of bulk proteins and valuable proteins from waste streams. Examples include recovery and purification of alkaline proteases from whole broths of *B. Lichenformis* by aqueous two-phase extraction using PEG and Reppal 200 (a partly hydrolyzed hydroxypropyl derivative of starch), recovery of bulk intracellular proteins from waste yeast from brewing operations using PEG/potassium phosphate extractant and recovery of β-galactosidase or lactase from homogenized yeast using PEG 400/PEG 6000/potassium phosphate system.

Partitioning of cell debris. Aqueous two-phase extraction is well suited for the removal of cell debris from homogenates. Disruption of biomass yields the desired product usually as part of a viscous colloidal suspension containing organalle and cell wall fragments along with other particulate matter. The separation of solids from such a suspension by conventional solid–liquid separation methods is difficult. This separation may be achieved by contacting the cell homogenate with PEG 1550/ potassium phosphate extractant to partition the cell debris and particulates into the heavier raffinate phase.

Affinity partitioning. The selectivity and yield of aqueous two-phase polymer extractions can be enhanced significantly by the use of affinity partitioning based on molecular recognition of the desired protein or enzyme by a ligand covalently bound to one of the phase-forming polymers. Based on thermodynamic considerations, if the protein to be extracted has N number of identical but independent binding sites for the affinity ligand, then the partition coefficient K in the presence of an excess of affinity ligand is given by:

$$K = K_0 \frac{(1 + [L]/K_T)^N}{(1 + [L]/K_B K_L)} \tag{6.28}$$

where K_0 is the partition coefficient for the protein in the absence of the ligand, $[L]$ is the total ligand concentration, K_B and K_T are the dissociation constants of the protein-ligand complex and K_L is the partition coefficient for the free ligand. The binding constant for protein-ligand complex determines the selectivity of extraction. The stronger the binding of the protein to the ligand, the lower the ligand concentration required for selective extraction. The selectivity increases exponentially with an increase in the number of binding sites N on the protein molecule. Thus the number of binding sites determines the yield. Using the affinity extraction selective extractions of different enzymes from yeast, porcine muscle, formate dehydrogenase from a cell homogenate of *Candida boidinii* and β-interferon from a mammalian cell culture medium have been reported.

Extraction of the desired protein by aqueous two-phase affinity partitioning can be manipulated by various factors such as the nature of the affinity ligand, phase-forming polymers, temperature, pH, protein concentration and type of salt used and its concentration.

6.11 REVERSED MICELLAR EXTRACTION

Reversed micellar extraction of biopolymers is yet another emerging technique. Surfactants aggregate spontaneously in organic solvents to form nanometer scale reversed or inverted micelle. The polar groups of the surfactant are buried within the core of the reversed micelle while the hydrophobic tail groups of the surfactant extend into the surrounding organic solvent, which is non-miscible with water. The reversed micelles also contain significant quantities of water and hydrophilic solutes. Reverse micellar extraction of proteins depends on the unique solubilizing properties of the reverse micelle. The aqueous feed is brought into contact with the reversed micellar phase (formed by dissolving the surfactant in an organic solvent, which is non-miscible with water), which functions as an extracting solvent. The aggregates in the extracting solvent consist of polar core of water, which partitions the proteins differentially surrounded by a protecting or stabilizing shell of surfactant layer (see Figure 6.10)

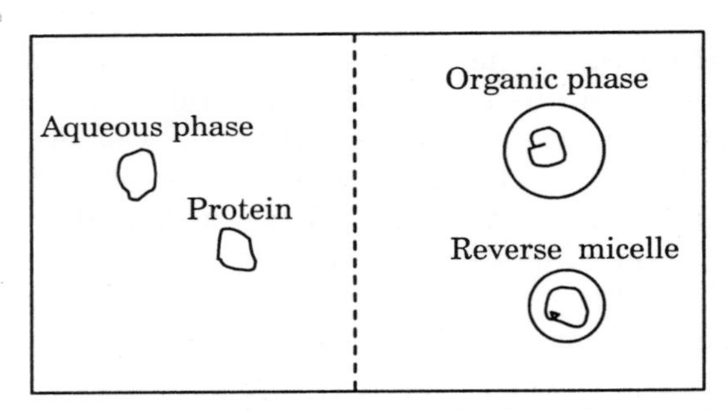

Figure 6.10 Reversed micellar extraction of the biopolymer.

Partitioning and extraction of proteins from the bulk aqueous phase into the reversed micellar organic solution is influenced by factors such as pH, ionic strength and the type of salt as well as organic solvents and surfactants. In general electrostatic interactions between the charged proteins and the surfactant polar head groups are responsible for solubilization of the proteins in reversed micelles. Any factor, which enhances the electrostatic interaction between the protein and the surfactant polar groups would facilitate the extraction of the protein. Thus for example, the solubility of a protein at pH less than its isoelectric pH (protein will have a net positive charge) is greater in an anionic surfactant and at pH greater than the isoelectric pH of the protein it would inhibit reversed micellar extraction. The converse is also true in that the solubility of the protein would be more at pH greater the isoelectric pH (protein would have a net negative charge) in a reversed micelle formed by a cationic surfactant.

Ionic strength plays a role in the solubilization process. Higher salt concentrations screen the electrostatic interactions and thereby decrease protein solubilization in reversed micelle. In addition, higher ionic strength results in the formation of small reversed micelles and the solubilization of the protein is reduced due to size exclusion effect, in which protein is less soluble in smaller reversed micelle. Further increased ionic strength brings about salting-out of the protein from the micellar phase.

Increase in surfactant concentration increases the number of reversed micelles and hence increases the solubilization of proteins. Reversed micelles formed by ionic surfactants have been found to be too small to accommodate large proteins while non-ionic surfactants form large sized micelles capable of dissolving a wide range of proteins.

Affinity partitioning of proteins is possible in reversed micellar extraction. The surfactant with a polar head group involving a specific substrate or inhibitor is chosen to extract the target enzyme selectively.

6.12 SUPERCRITICAL FLUID EXTRACTION

Supercritical fluid extraction is gaining importance particularly for extracting highly labile bioproducts such as food aroma components and flavours and bioproducts. The technique uses supercritical fluid carbon dioxide as the solvent.

6.12.1 Supercritical Fluid Solvents and their Properties

The critical temperature of a substance is defined as the temperature above which a distinct liquid phase cannot exist regardless of the pressure. The vapour pressure of the substance at its critical temperature is called the critical pressure. Alternatively it may be defined as the pressure required to liquefy a gas at its critical temperature. At temperatures and pressures above but close to the critical temperature and critical pressure (close to its critical point) a substance exists as a supercritical fluid.

Some of the potential extracting solvents include carbon dioxide, nitrous oxide, sulphur dioxide, ammonia, ethane, propane, butane, pentane and ethylene. Of these carbon dioxide finds wide use as the solvent because of its advantages of being non-toxic and non-hazardous nature.

The transport properties of a solvent such as density, viscosity and diffusivity as well as the thermophysical properties of a solvent such as heat capacity and latent heat of vapourization are of importance in extraction process. Supercritical fluids combine the high solvating characteristics of liquids with the low viscosity and high penetrating ability of gases. Supercritical fluids have physical properties such as density, viscosity and diffusivity and other properties that are intermediate between those of the substance in its gaseous and liquid states as shown in Table 6.3. Because of their low viscosity and greater diffusivity compared to liquids, supercritical fluids diffuse more rapidly into a sample and even penetrate solid samples. The solvent strength of a supercritical fluid can be controlled easily by increasing the pressure on the fluid. Increase in pressure increases the density of the fluid and thereby its solvent strength. Near the critical point small changes in pressure create large changes in the density of the supercritical fluid.

The latent heat of vapourization of the solvent decreases rapidly in the supercritical fluid region and becomes zero at the critical point as there is no phase transition involved. The heat capacity of the supercritical fluid is several times greater than that of the normal liquid.

TABLE 6.3 Physical Properties of Gases, Liquids and Supercritical Fluids

Property	*Gas (STP)*	*Supercritical fluid*	*Liquid*
Density (g cm^{-3})	$(0.6–2.0) \times 10^{-3}$	0.2–0.5	0.6–2.0
Viscosity (g cm^{-1}s^{-1})	$(1–3) \times 10^{-4}$	$(1–3) \times 10^{-4}$	$(0.2–3) \times 10^{-2}$
Diffusion coefficient (cm^2 s^{-1})	$(1–4) \times 10^{-1}$	$10^{-3} – 10^{-4}$	$(0.2–2) \times 10^{-5}$

6.12.2 Phase Diagram

The importance of the physical properties, particularly density of a supercritical fluid as a function of temperature and pressure is best described with the help of a phase diagram 6.11a. The pressure–temperature phase diagram of carbon dioxide shown in Figure 6.11b indicates the supercritical region of interest for use in extraction.

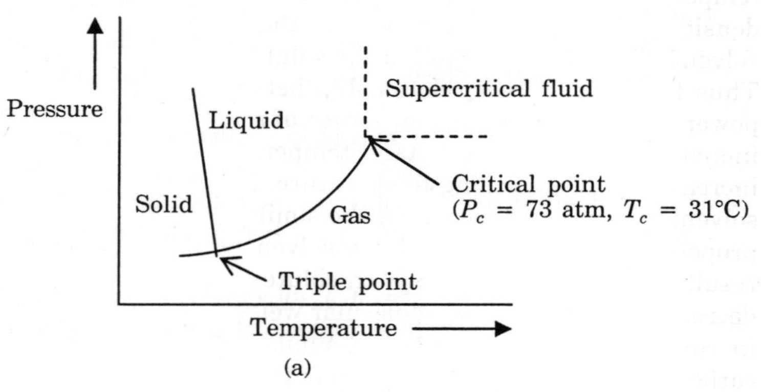

Figure 6.11 (Cont.).

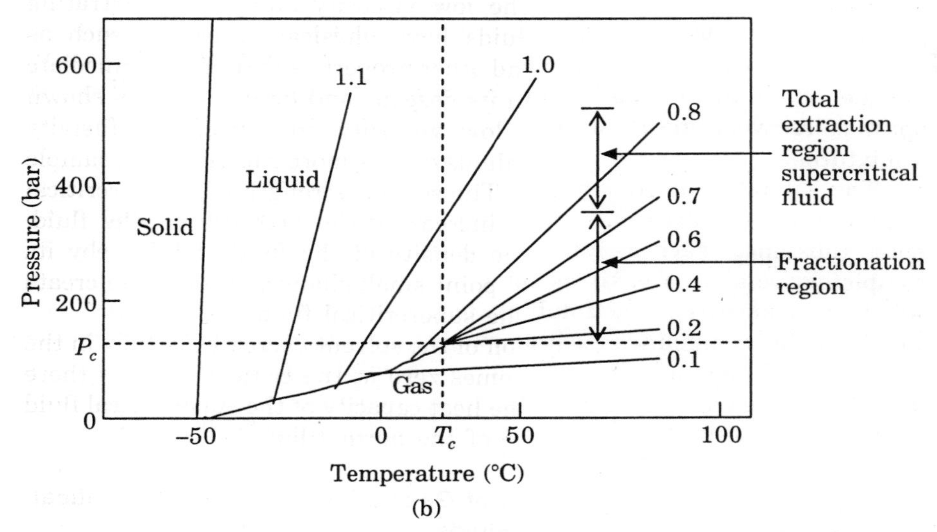

Figure 6.11 (a) Pressure-temperature phase diagram of carbon dioxide. (b) Supercritical region useful for extraction.

The diagram shows the pressure-temperature relationship at different density values as lines. Supercritical fluid extraction uses the total extraction region (density in the range of 0.6 < 0.9) and the fractionation region (density in the range of 0.2 < 0.6).

6.12.3 Supercritical Fluid Extraction Principles

The principle involved depends on the solvent power (the ability of the solvent to extract a particular component in a mixture) of the supercritical fluid. The solvent power is highly dependent on the density of the supercritical fluid, which in turn depends on the pressure and temperature. The close proximity of molecules in a liquid phase (greater density) imparts solvent power as the intermolecular forces allow the solvent molecules to surround the solute molecule and hold it in solution. Thus there is a direct relationship between the density and the solvent power. In general, the solvent power of a supercritical fluid increases with increasing density at constant temperature and at constant density it increases with increasing temperature. In the case of carbon dioxide, the solvent power also depends on the similarity of the physical and chemical properties of the solute and the solvent carbon dioxide, close similarity resulting in high solubility. With solutes of comparable polarity, volatility decreases with increasing molecular weight of the solute and the solubility in carbon dioxide also shows a similar trend. As the solvent power of carbon dioxide increases, the percentage solubility of a solute increases; the range of extractable solutes also increases and hence at high solvent power selectivity is low. In contrast, at low solvent power carbon dioxide exhibits high selectivity in dissolving solutes. Thus, supercritical carbon dioxide offers the possibility of tailoring the extracting conditions to optimize recovery and purification of the desired solute components from a mixture.

The solubility of organic compounds in supercritical carbon dioxide depends on the polarity and the molecular weight of the compounds. Non-polar substances such as aliphatic hydrocarbons and monoterpenes are completely soluble. Weakly polar compounds such as triglycerides are soluble to some extent, the solubility decreasing with increasing molecular weight. Polar compounds such as carboxylic acids and water are soluble to a negligible extent. The solubility of a solute in supercritical carbon dioxide may be enhanced by the addition of a co-solvent or entrainer. The addition of a co-solvent modifies the thermodynamic affinity between the solute and the extractant favourably.

6.12.4 Advantages of Supercritical Fluid Extraction

The technique of supercritical fluid extraction has several advantages.

- It offers a high degree of control over selective extraction of the desired component from a complex mixture.
- The method is economical and safe as it eliminates the need for large amounts of organic solvents.
- Recovery rates are higher than conventional liquid–liquid and liquid–solid extractions.
- The solvent carbon dioxide does not contaminate the final product as any residual solvent volatilizes off.
- The solvent is completely non-toxic to almost all bioproducts.
- The transport properties of the supercritical fluid facilitate better extraction and phase separation. Thus, the large density difference between supercritical fluid and the aqueous feed solution as well as the low viscosity of the solvent result in better mass transfer during extraction and improved phase separation after extraction. The droplet behaviour with a high degree of interfacial shear between the droplets and the continuous phase gives rise to improved mass transfer coefficient as well as improved phase separation. The diffusion coefficient of supercritical fluid is greater than that of a conventional liquid and thereby the kinetics of mass transfer involving diffusion limitation in the extract phase improves further.
- The unusual thermophysical properties of supercritical fluids are also advantageous during extraction. The low latent heat of

vapourization of the solvent allows its separation from the extracted product with low energy utilization. The heat capacity of the supercritical fluid is several times that of a normal liquid and hence heat transfer efficiency would be greater during product recovery.

- The low critical point of carbon dioxide allows it to be used for extracting heat-labile products.
- A variety of samples such as solids, semisolids and liquids of different chemical nature can be handled.

6.12.5 Extraction Process

Extraction using supercritical fluid can be potentially used for liquid–liquid as well as solid–liquid extraction (see Figure 6.12). In liquid–liquid

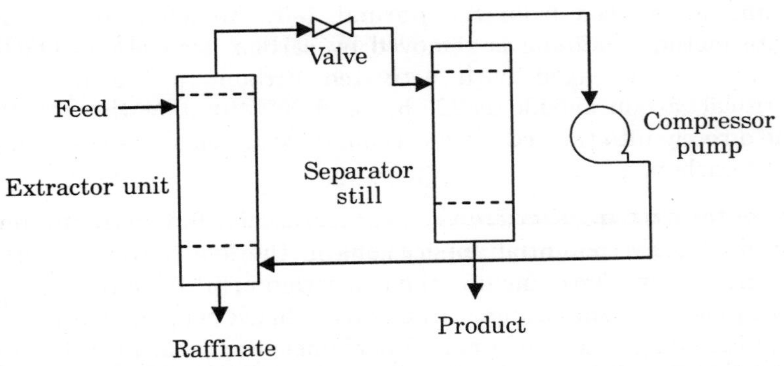

Figure 6.12 Schematic diagram of supercritical fluid extraction unit.

extraction, the aqueous feed is pumped into the extraction unit and contacted with the supercritical solvent at high pressures. The extract phase is brought to a lower pressure in a separator still to separate and recover the product from the solvent. The solvent is compressed and recycled. In solid–liquid extraction the supercritical solvent is pumped through a bed of solid feed to leach the desired solute. The extract phase is pumped to a separator still held at lower pressures to precipitate out the solid product. The solute-lean fluid is then compressed and recycled.

6.12.6 Applications

Extraction of bitter flavour from hops. Hops are dried flowers of the hop plant and are used to impart bitter taste to beer. The non-volatile soft resins of the hops contain the major active ingredients of alpha acids, which are desirable as well as the harsh bitter taste imparting beta acids. Organic solvents extract the desirable as well as undesirable components without discrimination. Supercritical fluid extraction using carbon dioxide allows the extraction of the soft resins particularly the alpha acids in their natural state. The extract is quite stable and can be stored for considerable period of time. Carbon dioxide, at a pressure of 58 bar and a temperature of 7°C, is pumped through a bed of the hops in a column to extract maximum amount of the alpha acids. The extract is pumped to a separator and carbon dioxide is evaporated at 40°C leaving the solid non-volatile products at the bottom of the vessel. The gas stream is decontaminated by carbon adsorption and compressed for recycling.

Decaffeination of coffee. The green coffee beans are soaked in water for some time and extracted with carbon dioxide at 160–200 bar and 90°C. The caffeine diffuses into supercritical carbon dioxide and is transferred to aqueous phase by washing with hot water. The caffeine is removed from the aqueous solution by distillation. The caffeine content in the beans is reduced from the normal 1–3% to about 0.04%. In an alternate method, caffeine is removed by carbon adsorption. The green coffee beans are mixed with activated carbon in the presence of supercritical carbon dioxide at 220 bar and 90°C to allow the diffusion of caffeine directly into the activated carbon. Sieving separates the beans and activated carbon.

Other potential applications. Supercritical fluid extraction using carbon dioxide has potential applications in the food flavours, perfumes and oil industries. These include the extraction of (i) flavours from fruits such as apple, pear and orange; (ii) oils from black pepper, almond, lemon peel and lilac for use as flavours and perfumes; (iii) β-carotene from plant materials; (iv) oils from seeds of sunflower, soya bean and rape; (v) monoglycerides from vegetable oils for use as emulsifiers and (vi) common organic chemicals such as alcohols, ketones, carboxylic acids, esters etc. from aqueous media.

Exercises

1. The partition coefficient value of an organic acid in organic solvent–water system is 2.7. (a) Calculate the volume of organic solvent required to extract 99% of the acid from 50 ml of aqueous solution. (b) How many extractions with 50 ml of organic solvent would be required to extract 99% of the acid?

Solution

(a) Extraction of 99% of acid leaves behind 1% in the aqueous phase at the end of single extraction.

$$C_1 = \frac{C_0}{[1 + K(V_L/V_H)]^1}$$

$$1 = \frac{100}{[1 + 2.7(V_L/50)]}$$

$$V_L = 95 \text{ ml}$$

(b) For multiple extractions using 50 ml of organic solvent to achieve 99% extraction:

$$1.0 = \frac{100}{[1 + 2.7(50/50)]^n}$$

$$\frac{1.0}{100} = \left(\frac{1}{3.7}\right)^n$$

$$\log 0.01 = n \log 0.27$$

$$n = \mathbf{3.52} \text{ (rounded off to 4)}$$

Multiple extractions involving 50 ml of organic solvent requires 4 stages to remove 99% of the acid from the aqueous feed.

Thus multi-stage extractions are more efficient in that the volume of organic solvent required is less compared to single stage extraction.

2. An antibiotic exhibits a K value of 10 in organic solvent-water system. If the aqueous feed has 25 mg of the solute (a) how much could be extracted with an equal volume of organic solvent? (b) What percentage of the antibiotic will be extracted if four equal portions of same volume of organic solvent is used?

Solution

(a) For single stage extraction using the same volumes of V_L and V_H

$$C_1 = \frac{25}{[1 + 10(V_L/V_H)]^1}$$

$$= \frac{25}{11} = 2.27 \text{ mg remains in the raffinate}$$

Percentage extracted in single stage = (25 – 2.27)/25 = **91%**

(b) For multi-stage extraction, 4 equal portions of the same volume are used.

$$V_L = \frac{V_L}{4}$$

Percentage remaining after four stages of extraction

$$= 100/[(1 + 10(V_L/4))]^4$$

$$= 100/(1 + 2.5)^4 = 0.0286\%$$

Percentage extracted after four stages of extraction

$$= 100 - 0.0286 = 99.97\%$$

Once again it is clear that multi-stage extraction is more efficient.

3. Two weak acids HA and HB have dissociation constant K_a values of 1.2×10^{-8} and 1.8×10^{-4}, respectively. The intrinsic distribution coefficients (K_i) of HA and HB between water and an organic solvent are 9.8 and 6.3 respectively. What should be the extraction pH at which 99% of HA will be extracted into the organic solvent while leaving behind 99% of HB in the aqueous phase?

Solution

From equation 6.5

(i) For retaining 99% of HB in the aqueous phase, the distribution ratio K is taken as 0.01 for HB and then pH is calculated.

$$K = 0.01 = \frac{6.3}{1 + 1.8 \times 10^{-4}/[H_3O^+]}$$

$$0.01 = \frac{6.3[H_3O^+]}{[H_3O^+] + 1.8 \times 10^{-4}}$$

$$[H_3O^+] = \frac{1.8 \times 10^{-6}}{6.29} = 2.86 \times 10^{-7} \text{ or pH} = \mathbf{6.54}$$

(ii) By carrying out the extraction at pH 6.54, the distribution ratio K for HA is calculated.

$$K = \frac{9.8 \times 1}{(1 + 1.2 \times 10^{-8}/2.86 \times 10^{-7})} = \mathbf{9.4}$$

Questions

1. What are the advantages of liquid–liquid extraction?
2. Explain the principles of extraction.
3. Write a note on the different extraction processes.

4. Discuss the operating principles and analysis of single and multiple stage batch extractions.

5. How are continuous stage extractions carried out?

6. What are aqueous biphasic systems? Give examples. How are they useful?

7. Give an account of the theoretical principles and steps involved in the aqueous two-phase extraction of an enzyme.

8. How is reverse micellar extraction of a biopolymer carried out?

9. What is a supercritical fluid? What are its characteristics?

10. Discuss the principles of supercritical fluid extraction.

CHAPTER

7

Membrane Separation Processes

7.1 MERITS OF THE PROCESS

Membrane separation consists of different processes operating on a variety of physical principles and applicable to a wide range of separations of miscible components. These methods yield only a more concentrated liquid stream than the feed. Membrane separation processes have several advantages. These include:

- The method is a low energy alternative to evaporation in concentrating a dilute feed, particularly when the desired material is thermally labile or when the desired component is a clear liquid.
- The chemical and mechanical stresses on the product are minimal and since no phase change is involved the energy requirement is modest.
- Product concentration and purification can be achieved in a single step and the selectivity towards the desired product is good.
- The method can easily be scaled up.

In bioprocess industry, membrane separation is widely used because of the mild operating conditions and low energy requirements in the recovery of lactose from whey, separation of immiscible components such as recovery of extracellular products (e.g., proteins, enzymes, etc.) and biomass from fermentation broths. Membrane separation process cannot be used for feeds containing a high concentration of low molecular weight components because of high osmotic pressure or when the feed has high solid content (>25% w/v) because of pumping problems.

7.2 CLASSIFICATION OF MEMBRANE SEPARATION PROCESSES

The common factor in all the membrane separation processes is the physical arrangement of the process in which a membrane acts as a semi-permeable barrier between two fluid phases made up of two liquids (or two gases or a liquid and a gas). The membrane prevents actual hydrodynamic flow of the two phases. The semi-permeable membrane differentiates between solutes of different sizes. Separation occurs as the membrane controls the rate of movement of various components through selective transport, allowing some components to pass through while retaining others.

Membrane separation processes may be classified broadly into three categories on the basis of the driving force, which facilitates mass transfer across the membrane. In the first category the driving force is hydrostatic pressure and the separation methods include (i) microfiltration (MF) (ii) ultrafiltration (UF) and (iii) reverse osmosis (RO) or hyperfiltration. In the second category, the driving force is concentration and the method is called dialysis. Electrodialysis belongs to the third category wherein the driving force is an applied electric field. A schematic representation of the process flow diagrams of membrane separation processes is shown in Figure 7.1.

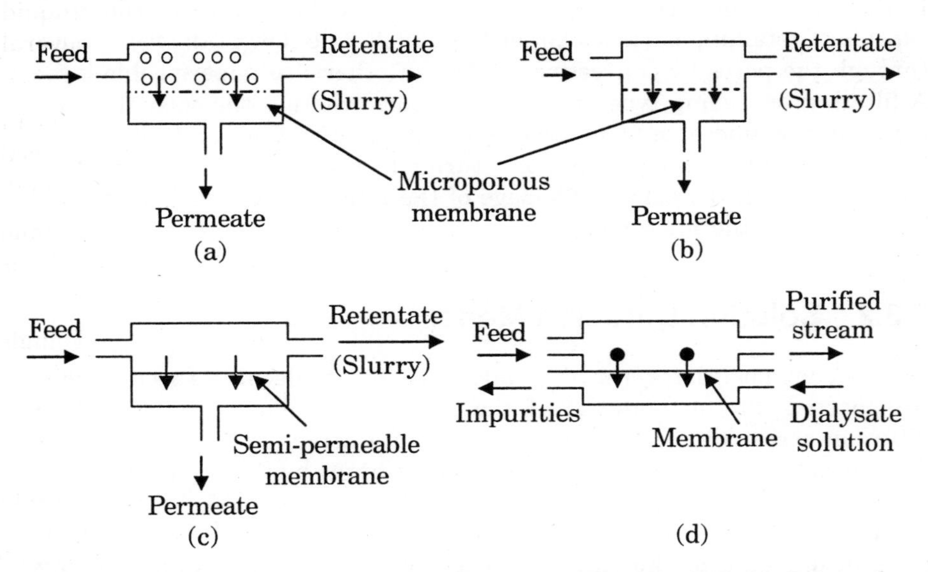

Figure 7.1 Process flow diagrams in (a) MF (b) UF (c) RO and (d) dialysis.

The characteristics of the membrane separation processes are summarized in Table 7.1.

TABLE 7.1 Characteristics of Membrane Separation Processes

		Characteristic features	
Process	*Driving force*	*Membrane pore size*	*Separation mechanism*
MF	Pressure 0.1–1 bar	0.02–10 μm	Sieving
UF	-do- 2–10 bar	0.001–0.02 μm	Sieving
RO	-do- 10–100 bar	Non-porous	Solution diffusion
Dialysis	Concentration difference	1–3 nm	Sieving and diffusion
Electrodialysis	Electrical potential	Mol.wt < 200	Ion migration

7.3 THEORETICAL MODELS FOR MEMBRANE PROCESSES

Two basic models of mass transport across the membrane have been put forward to explain the selectivity exhibited by the membrane separation processes. These include (i) capillary flow model and (ii) solution-diffusion model.

7.3.1 Capillary Flow Model

In this model, the membrane is considered to be 'loose' and microporous, capable of retaining particles larger than 10 Å. The flow of the feed occurs through the pores by convective flow in an otherwise impermeable layer. A filtering or sieving type mechanism occurs wherein the solvent moves through the micropores in essentially a viscous flow and the solute molecules small enough to pass through the pores are carried by convection with the solvent. Passage of the larger molecules is prevented by the size of the micropores.

7.3.2 Solution Diffusion Model

This model postulates the dissolution of the molecular species being transported, in the material of the membrane followed by molecular diffusion across the barrier obeying Fick's diffusion. The driving force is the concentration gradient set up in the membrane by the applied pressure difference. The membrane is 'tight' in that it is capable of retaining solutes of about 10 Å in size or less. Since both the solubilities and the rates of diffusion of the various molecular species will be different, this model explains the selectivity of a reverse osmosis membrane to different components in solution.

Both the mechanisms probably occur in membrane transport, the solution-diffusion mechanism predominating in reverse osmosis and capillary flow in microfiltration and ultrafiltration. It is apparent that chemical nature or molecular structure of the membrane appears to be more important in reverse osmosis than in the other two membrane separations. Microfiltration and ultrafiltration may be regarded as an extension of filtration operation to molecular size range with the membrane functioning as a sieve. The presence of relatively large-sized pores does not contribute any selectivity in the separation process, the separation being based only on molecular size.

7.4 RETENTION COEFFICIENT OR REJECTION COEFFICIENT

The separating ability of a membrane in pressure driven processes of MF, UF and RO is explained in terms of retention or rejection coefficient which is defined as given by Eq. (7.1).

$$R = \frac{C_m - C_p}{C_m} \tag{7.1}$$

where R is the theoretical retention coefficient and C_m and C_p represent the concentrations of the solute at the membrane surface and in the permeate respectively. The actual observed retention coefficient, R' is related to the concentrations of the solute in the bulk phase, C_b and permeate C_p as given by Eq. (7.2).

$$R' = \frac{C_b - C_p}{C_b} \tag{7.2}$$

The theoretical and observed retention coefficients are related as given by Eq. (7.3).

$$R' = 1 - (1 - R)\frac{C_m}{C_b} \tag{7.3}$$

Due to concentration polarization at the membrane surface, the ratio $C_m/C_b > 1$ and the observed retention coefficient will be less than the theoretical or true retention coefficient. Concentration polarization increases solute leakage through the membrane particularly for RO process while the build-up of solute particles on the membrane surface will function as a barrier and reduce solute leakage through the membrane in the case of MF and UF.

7.5 FACTORS AFFECTING THE SEPARATION PROCESSES

Two factors affect the selectivity and the flux or permeation rate through the membrane severely affecting the overall performance of the separation process. These include (i) concentration polarization at the membrane surface, which is a short-term effect and also reversible and (ii) fouling of the membrane, which is a long-term one and irreversible.

7.5.1 Concentration Polarization

The non-permeating species is carried towards the membrane by the convective flux of the feed but the species remains on the upstream side. Its concentration increases gradually at the membrane surface and ultimately becomes greater than its concentration in the bulk liquid thereby setting up concentration polarization at the membrane surface. It sets in also due to different rates of transport of various species.

Concentration polarization augments osmotic pressure and reduces the flux through the membrane and may affect membrane separation characteristics. The phenomenon is important, particularly in ultrafiltration.

The steady state solute concentration profile in the boundary layer close to the membrane surface is shown in Figure 7.2.

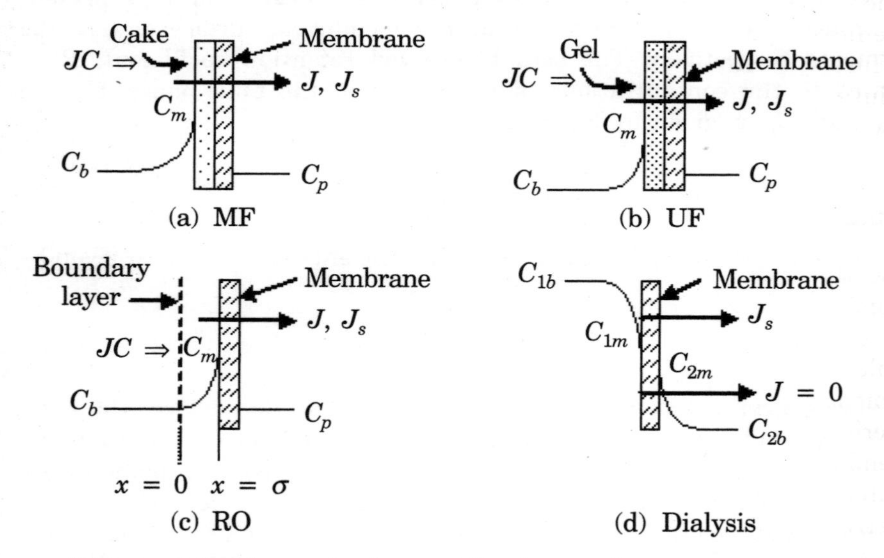

Figure 7.2 Concentration profiles in membrane processes.

The solute concentration at the membrane may be calculated from a mass balance on the solute:

Rate of convection towards the membrane = Rate of diffusion back into the bulk liquid + Rate of permeation

$$J_s C = \frac{D\,dC}{dx} + J_s C_p \tag{7.4}$$

$$\frac{dC}{C - C_p} = \frac{J_s}{D}\, dx \tag{7.5}$$

Integrating from $x = 0$ and $C = C_b$ to $x = \sigma$ (σ = thickness of the bound layer) and $C = C_m$; and substituting for (D/σ) by k', a mass transfer coefficient, the concentration of the solute in the membrane surface is given by:

$$C_m = C_p + (C_b - C_p) \exp\left(\frac{J_s}{k'}\right) \tag{7.6}$$

or

$$J_s = k' \ln\left(\frac{C_m - C_p}{C_b - C_p}\right) \tag{7.7}$$

If a negligible amount of solute passes through the membrane, C_p can be taken as zero, giving:

$$C_m = C_b \exp\left(\frac{J_s}{k'}\right) \quad \text{or} \quad J_s = k' \ln\left(\frac{C_m}{C_b}\right) \tag{7.8}$$

The ratio C_m/C_b increases if the flux, J_s is high or the mass transfer coefficient, k' is low. Low k' values may be due to low values of the diffusivity, D, due to high molecular weight or due to high viscosity of the solution, or large values of the thickness of the boundary layer σ due to low turbulence. The value of k' has to be determined empirically.

7.5.2 Fouling

The flux through the membrane decreases slowly with time in all the membrane processes due to fouling caused by a variety of factors such as (i) slime formation, (ii) microbial growth, (iii) deposition of macromolecules (particularly in UF), (iv) colloid deposition and (v) physical compaction of the membrane (particularly in RO due to high pressure operation). Fouling is irreversible and necessitates the replacement of the membrane. Fouling can be inhibited by careful selection of the membrane material (e.g. hydrophilic surface is less prone to fouling by proteins), pretreatment of the feed (such as pH adjustment or precipitation to remove salt), frequent cleaning of the membrane with chemicals and backflushing with permeate.

The problems associated with concentration polarization and fouling is overcome by cross-flow filtration instead of the conventional dead-end filtration. In the conventional filtration, the feed is pumped perpendicularly to the filter medium. As the fluid passes through the filter, a concentrated solution or gel of non-permeating species is formed on the upstream side of the filter resulting in concentration polarization. Increasing the velocity at the membrane surface creates turbulence thereby decreasing the thickness of the concentration boundary layer and delaying the onset of concentration polarization. However, a better strategy is to adopt cross flow filtration (vide Figure 7.3) instead of the conventional dead-end filtration. In cross-flow filtration, the feed fluid is pumped tangentially across the filter medium (and perpendicular to the flux through the membrane) to avoid concentration polarization. Other advantages of this technique are that the membrane separation characteristics are not affected and the flux is not reduced with filtration time.

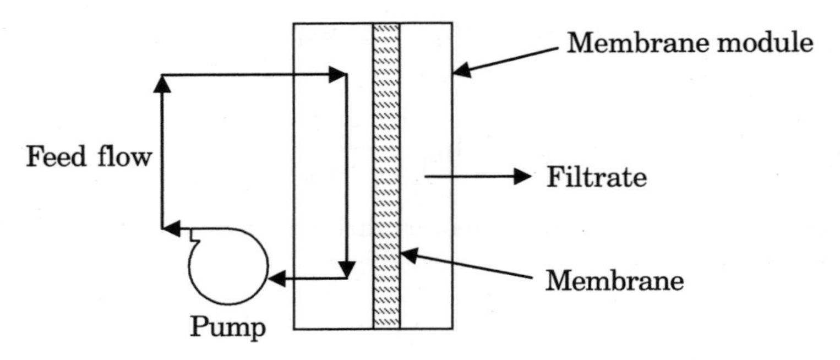

Figure 7.3 Cross-flow filtration.

7.6 OPERATIONAL REQUIREMENT OF MEMBRANES

The operational requirements of the membranes of the pressure driven processes are similar in that they should have (i) selectivity and high separation efficiency, (ii) high permeate flux rate, (iii) mechanical/physical strength to withstand the high pressure operation without elongation of the pores, (iv) durability and consistency of performance over prolonged periods, (v) resistance to corrosion and (vi) ease of fabrication in appropriate shape. In addition, the membrane material should be of low cost and readily available. The operational requirements of a reverse osmosis membrane are more stringent in that the membrane should be able to discriminate between the solvent molecules and the low molecular weight solutes and ions and allow the flow of only the solvent. In aqueous solutions of simple electrolytes and low molecular weight solutes (MW < 500 Da), the molecular dimensions of the solute and solvent are comparable and the osmotic pressure developed is quite large. Hence the reverse osmosis membrane should be capable to withstand high pressure operation in the range of 50–60 atmospheres.

7.7 STRUCTURE OF MEMBRANES

The semipermeable membranes used in RO and UF processes have generally two phases. A thin (0.5–10 μm) dense layer of material with microporous structure covers the top of the membrane and is supported by a thicker layer (50–125 μm) of relatively macroporous material. The top layer is responsible for the basic separation characteristics while the bottom layer gives the strength to the membrane. The top layer and the porous support are made in a single casting process to give a membrane of 0.1–0.2 mm in thickness. The membrane is supported on a rigid, porous backing structure.

7.8 PREPARATION OF MEMBRANES

The preparation of RO and UF membranes involves various considerations.

The choice of membrane material is the most important factor. Synthetic membranes based on cellulose acetate, cellulose phthalate, polyamides, polyacrylonitrile, polyethylene and polytetrafluoroethylene are available. In the first step, the base material is dissolved in a suitable solvent with necessary additives to give a homogenous solution of desired viscosity. In the second step, a film is cast by spreading the solution on a clear, dry glass plate or drawn as a hollow tube with proper thickness control. In the third step, the cast film is maintained in a controlled atmosphere to allow evaporation of the solvent. The evaporation time and atmosphere are very important in terms of basic quality of the membrane. In the fourth step of gelation, the membrane is dipped in a water bath at precisely controlled temperature in the range 0–4°C, to leach out the solvent and the additives slowly thereby creating micropores in the membrane. In the final step, annealing is carried out by heating the membrane in a water bath at 70–90°C to allow the shrinkage of pores formed during gelation. Annealing is carefully controlled to get optimum performance of the membrane with reference to separation efficiency and permeation rate. High temperature annealing results in higher separation efficiency but the permeation rate decreases considerably.

The conditions employed for the preparation of cellulose acetate membrane for a typical desalination process are as follows. The casting solution contains cellulose acetate (25%), acetone (45%) and formamide (30%). In addition small quantities of (1–5 g) of magnesium perchlorate is also used. The evaporation time is about 2–3 minutes at 20–25°C, gelation is at 2°C for 5 minutes and annealing is at 80°C. The solutes most effectively excluded by the cellulose acetate membrane are the salts such as sodium chloride, sodium bromide, calcium chloride and sodium sulphate. Sucrose and tetra alkyl ammonium salts are also excluded. The main limitation of cellulose acetate membrane is that it can be used only for aqueous solution and at temperatures below 60°C.

Medium porosity polypropylene membranes of thickness of about 0.03 mm are made by drawing nonporous films of the polymer. The polymer tears to give small pores when they are annealed into the membrane structure. The membranes are hydrophobic and hence have to be primed with alcohol-water mixtures before using for aqueous solutions.

Membranes with most monodisperse pores and of very low porosity are made by exposing nonporous films of mica or polycarbonate to α radiation. The radiation tracks in the film are then etched away with hydrofluoric acid.

Composite membranes have been developed by the superposition of several layers of metallic oxides (usually zirconia) on a calcined carbon support. The composite membranes have several advantages such as capability to withstand high stress during operation, high shock resistance, suitability to undergo steam sterilization and capability to handle hot process fluids, high concentrations and high viscosity of the feed.

7.9 EQUIPMENT

The components of a typical membrane separation plant are shown in Figure 7.4.

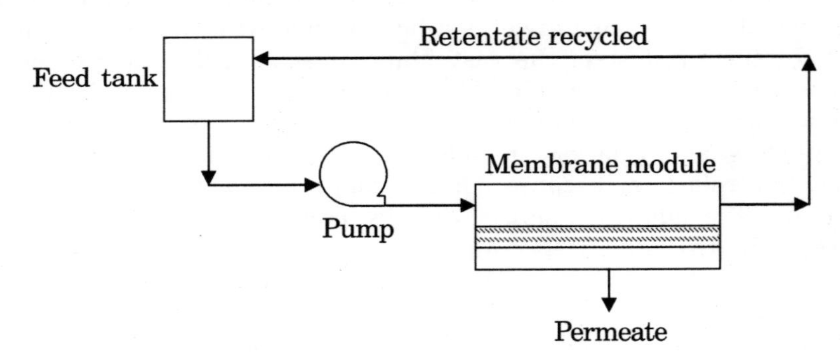

Figure 7.4 Membrane separation unit.

These include (i) reservoirs for the feed, permeate and reject solutions, (ii) high pressure pump (piston, diaphragm or centrifugal type), (iii) membrane modules of suitable configuration and (iv) high pressure regulators and gauges. The feed is to be pretreated by removing suspended particulates and turbidity to avoid fouling of the membrane, which reduces the separation efficiency and permeate flow.

Different membrane modules have been designed on the basis of criteria which include (i) a high membrane surface–volume ratio to minimize space requirement and capital cost, (ii) an adequate structural support to allow the thin membrane to withstand the required operating pressure, (iii) a low pressure drop on the concentrate side of the membrane to maintain the driving force for permeation, (iv) turbulence on the concentrate side to dissipate concentration polarization and minimize fouling and (v) provision for back flushing and replacement of membrane. The different membrane module configurations include (i) flat sheet membrane, (ii) spiral-wound membrane, (iii) tubular and (iv) hollow fiber modules (see Figure 7.5).

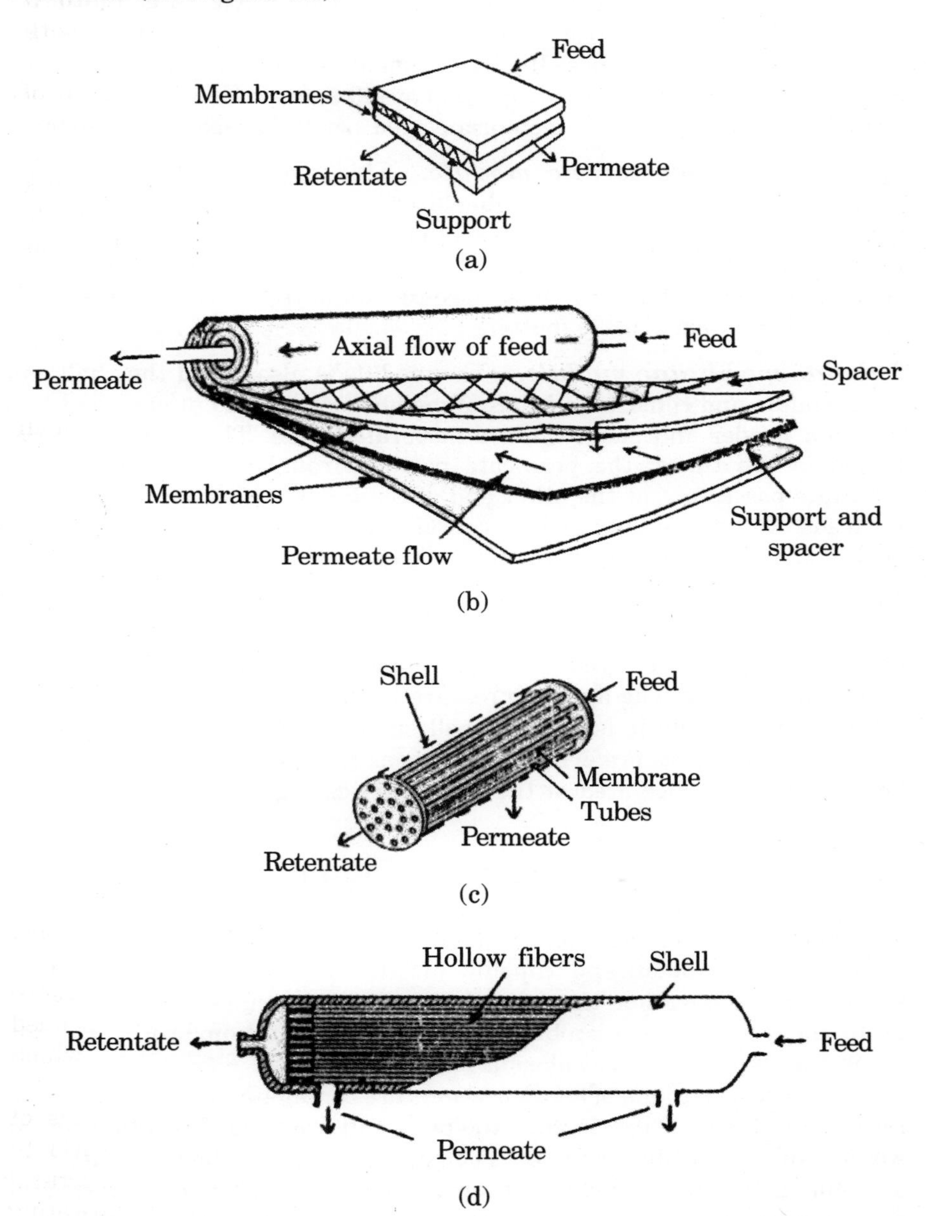

Figure 7.5 Membrane modules (a) flat sheet (b) spiral wound, (c) shell and tube (d) hollow fiber.

Flat sheet membrane module. Flat sheet membrane modules are fabricated by stacking several flat sheets of membrane as a multilayer sandwich in a plate-and-frame filter press type arrangement. The module is easy to fabricate and use. Its arrangement is compact and the module can withstand high pressure drop up to 30–40 kg/cm^2. However, the major drawback is its very small membrane area per unit separator volume.

Spiral-wound membrane module. This module is fabricated by winding a membrane sheet in double layer into a spiral. The feed is passed axially or spirally over the outer side of the double layer. The permeate moves spirally to a pipe located at the center of the spiral. The spiral wound membrane has a high surface-to-volume ratio. It requires clean feeds free from particulate matter.

Tubular membrane module. The module is also called the shell and tube module and consists of tubes of membranes bonded at each end to a common header and packed into a perforated shell. The feed enters the lumen of the tubes, the permeate passes through the wall while the retentate passes out at the other end of the tubes. This module finds use for feeds of high viscosity or feed containing particulate matter such as a feed of tomato puree which requires to be concentrated.

Hollow fiber membrane module. The module consists of hollow capillary fibers of membrane packed into a shell-and-tube arrangement to provide a high surface area per unit volume. The feed must be free from particulate matter. The applied pressure is on the inside of the fibers for ultrafiltration while it is on the shell side for reverse osmosis. Hollow fiber configuration is preferred in dialysis as the driving force is small and hence a large surface area of the membrane is required.

7.10 MICROFILTRATION OR CROSS-FLOW FILTRATION

Microfiltration resembles conventional filtration, the filter medium being a microporous membrane capable of filtering particulate matter or suspended solids rather than dissolved solute molecules. The method is useful for harvesting microbial cells from fermentation broths and for separating blood cells and plasma from whole blood.

Three basic designs of configurations are commonly used for microfiltration in industry. These include the plate and frame configuration, spiral wound and hollow fiber modules. The important part of the microfiltration assembly is the thin and porous filter medium or membrane supported on a rigid support. Synthetic organic polymer (polypropylene or PTFE) as well as inorganic (alumina or zirconia) membranes with well-defined pores with high thermal and chemical resistance are commercially available. The choice of the membrane material depends on parameters such as wettability, adsorption characteristics as well as chemical and mechanical stability. The membrane allows even macrosolutes but not colloids and suspended solid particles in the size range of 0.1–10 microns. The flow of the solvent and solutes through the membrane is primarily due to convective flow through the pores. The convective transport is pressure-driven, the operating pressure being relatively small in the range of 1 to 2 bar.

Microfiltration may be operated in normal flow (dead-end) mode or cross-flow mode. In the normal flow mode, the feed flow is perpendicular to the filter medium while in cross-flow mode the feed flow is tangential to the filter medium. In the normal flow mode, as filtration progresses, resistance to liquid flow increases due to cake formation on the upstream side of the filter medium causing a pressure drop. In most bioseparations, the cake is compressible causing a non-linear variation in the pressure drop. The flux (flow of feed per unit area) is given by the equation:

$$J = \frac{\Delta P}{r_m + r_c} \tag{7.9}$$

where r_m and r_c are the resistance of the membrane and of the cake to fluid flow and ΔP is the pressure drop. The pores in the filter medium are small and highly monodisperse compared to the medium used in conventional filtration and retains the larger particles but allows the liquid and smaller particles to permeate through. Since the pores are small, the resistance of the filter medium is quite high, i.e., it has a low Darcy's law of permeability. The resistance of the cake is given by:

$$r_c = \frac{\alpha w \Delta P^s V \eta}{A} \tag{7.10}$$

where α is the specific resistance of the cake, w is the slurry concentration (dry solids/m^3 of filtrate), s is the compressibility exponent (0 for incompressible cake and 1 for perfectly compressible cake), V is the total volume of the permeate, η is permeate viscosity and A is the area of the membrane. Substituting for r_c in the above equation:

$$J = \frac{\Delta P}{(\alpha w \Delta P^s V \eta / A) + r_m} \tag{7.11}$$

If $r_m << r_c$ then the total volume of the permeate that will pass through the membrane of area A is:

$$V = \frac{A\,\Delta P^{(1-s)}}{\alpha w \eta J} \tag{7.12}$$

If $s = 1$ then the flux through the membrane is independent of pressure drop. In many cases the α, w, η and the total flow rate through the membrane JA are constant.

A disadvantage of microfiltration is the frequent fouling of the membrane due to deposition of solids on the membrane which necessitates purging of the unit periodically. At some limiting pressure drop, the membrane needs to be cleaned or replaced. The total volume of the permeate that will pass through the membrane of area A at the time when the membrane needs to be replaced or cleaned can be calculated from Eq. (7.12). In general, increasing the area of the membrane twice decreases the flux by half, but the total volume of the filtrate will increase four times. The overall effect of increasing the area of the membrane is that it will reduce the membrane replacement cost by half.

The build-up of solids on the membrane can be minimized by installing a depth filter upstream of the microfiltration unit to pre-filter the particles. Periodical back washing and using cross flow of the feed tangential to the surface of the filter medium is also practiced. The large ratio of cross flow to flow through the filter minimizes the accumulation of solids on the filter medium and hence no cake is formed. This is an advantage over conventional filtration because maximum pressure drop in the latter is primarily due to cake formation. In addition complications occur in conventional filtration due to non-linear relationship between flow rate and pressure drop, particularly when the cake is compressible.

Another disadvantage of microfiltration is that concentrated slurry, and not a dry cake is obtained. The initial feed is usually a highly dilute suspension and as filtration progresses and liquid (permeate) is removed, the suspension becomes concentrated and its viscosity increases. Rapid pumping to sustain cross flow becomes gradually more and more difficult and hence the suspension is discharged even though it still contains a large amount of liquid. Hence microfiltration is followed by conventional filtration or centrifugation to separate the solids completely.

7.11 ULTRAFILTRATION

Ultrafiltration is operated at pressures in the range of 2 to 10 bar, the separation process occurring across a membrane, which discriminates solute molecules on the basis of their size. The method is useful in the separation of high molecular weight products such as polymers, proteins and colloidal materials from low molecular weight solutes. It is used in food industry to concentrate and clarify fruit juices. In dairy industry, it is used for the recovery of whey proteins during cheese manufacture. Ultrafiltration finds use to concentrate cell-free fermentation broths containing complex products such as monoclonal antibodies in pharmaceutical industry.

The ultrafiltration membrane made from polysulphone or other polymers is finely microporous and may be asymmetric. The pore size is smaller than that in microfiltration membrane and the pore area per unit surface is also less and hence the flux is also much less. Liquid flow through the membrane is by viscous flow through the pores due to the moderate applied pressure. The osmotic pressure is negligible because of the high molecular weights of the solutes.

The flux J in ultrafiltration is reduced to a significant extent due to concentration polarization. The solute retention characteristics of the membrane are also affected due to concentration polarization. The increase in pressure drop seems to have no influence on the flux beyond certain limiting flux. This behavior has been attributed to changes in osmotic pressure due to concentration polarization or the formation of an incompressible gel layer. Consequently the equation for flux is given as:

$$J = k \ln\left(\frac{C_g}{C_b}\right) \tag{7.13}$$

where C_g and C_b represent the concentration of the solute in gel layer and the bulk feed respectively and k is a constant. J is independent of pressure drop across the membrane as shown in the above equation since all the terms in the right hand side are constant.

A mass balance for the solute undergoing ultrafiltration can be written as:

Rate of loss of solute = solute flow through permeate

$$-d\left(\frac{VC}{dt}\right) = JAC_p \tag{7.14}$$

or

$$-d\left(\frac{VC}{dt}\right) = JAC(1 - R) \tag{7.15}$$

where V is the retentate volume, C and C_p are the concentration of the solute in the retentate and permeate respectively, A is area of the membrane and R is the retention coefficient. If the retention coefficient for a solute $R = 1$, then VC will be a constant as per the above equation. However if $R < 1$ then:

$$-C\left(\frac{dV}{dt}\right) - V\left(\frac{dC}{dt}\right) = JAC\ (1 - R) \tag{7.16}$$

Replacing $(-dV/dt)$ by JA and V by $V_i - JA_t$ and integrating from $t = 0$, $C = C_i$ to $t = t$ and $C = C_f$ gives:

$$\frac{C_f}{C_i} = \left(\frac{V_i}{V_f}\right)^R \tag{7.17}$$

The amount of solute lost is then $V_iC_i - V_fC_f$.

The commercially available ultrafiltration membranes are characterized by the *nominal molecular weight cut-off* (NMWCO) derived from tests in which the rejection coefficient profile of a membrane to known molecular weight solutes is evaluated. The NMWCO represents the molecular weight for which the rejection coefficient is a fixed percentage (usually 90%). However this value may be significantly lower in real situation because of the fact that interactions in the vicinity of the membrane surface changes the separation characteristics of the membrane.

7.12 REVERSE OSMOSIS (RO) (HYPERFILTRATION)

Reverse osmosis uses membranes (pore size is in the range of 0.0001–0.001 μm) permeable to water but not dissolved salts of even low molecular weight. The nomenclature is based on the fact that the direction of normal osmotic flow of a solvent across a semi-permeable membrane is reversed due to an applied pressure which is greater than the osmotic pressure of the liquid feed. In normal osmotic process, the solvent diffuses through a membrane, separating the solution containing low molecular weight products and the pure solvent, from the solvent side to the solution side. In reverse osmosis, a reverse pressure difference (20–100 bar) imposed on the membrane causes the flow of the solvent from the solution side to the pure solvent side thereby concentrating the solution (by removal of solvent). The method is applicable to separate low molecular weight products such as salts (as in desalination of seawater), sugars or organic acids from aqueous solutions. Its use has been extended to food and dairy industries to concentrate fruit juices, vegetable juices and milk.

Nanofiltration is similar to reverse osmosis but the membranes are slightly more porous and can be used to separate molecules up to 500 Da.

7.12.1 Osmotic Effect

When an ideal semi-permeable membrane forms a barrier between a solution and its pure solvent, the solvent molecules pass spontaneously across the membrane from the solvent to the solution. In general, the solvent molecules pass from a region of lower concentration to a region of higher concentration. The driving force for the flow of solvent across the membrane is the difference in the chemical potential on the two sides of the membrane. This phenomenon is called osmosis (vide Figure 7.6a). The flow of the solvent across the membrane continues until the fluid pressure in the concentrated solution is sufficiently high to prevent the passage of further solvent molecules. At equilibrium, the chemical potentials on both sides of the membrane are equal and the fluid pressure is called the osmotic pressure of the solution (see Figure 7.6b).

The osmotic pressure is characteristic of a particular solvent-solute system and is proportional to the concentration of the solute and the absolute temperature. Van't Hoff showed that the relationship is similar

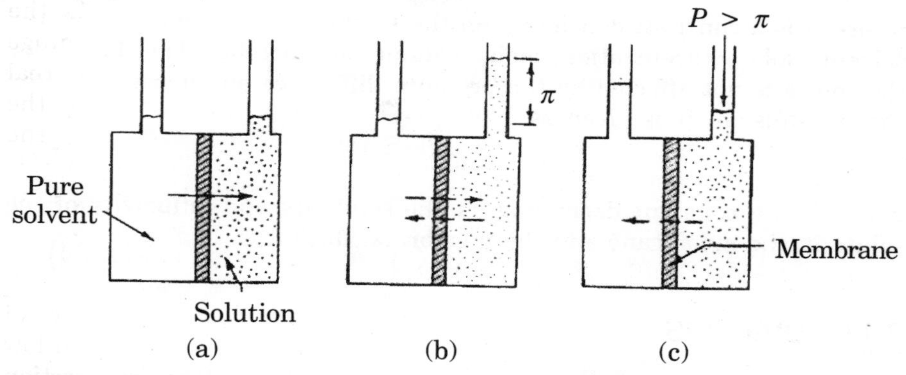

Figure 7.6 Schematic diagram of osmosis and reverse osmosis.
Source: C.J. Geankoplis, *Transport Processes and Unit Operations*, 3rd ed., Prentice-Hall of India, New Delhi, 2002.

to that of pressure of an ideal gas and for a dilute aqueous solution may be expressed as:

$$\pi = RTC \tag{7.18}$$

where π is the osmotic pressure, C is the concentration in number of moles of solute/m^3, R is the gas constant (in litre atm.) and T is the temperature in Kelvin. The equation is strictly valid for dilute solutions. For more concentrated solutions, the actual (measured) osmotic pressure is less compared to that calculated on the basis of the equation. The ratio of actual osmotic pressure to the calculated value is called the osmotic coefficient, ϕ, which is included as a correction factor. The value of the osmotic coefficient is unity in dilute solutions and decreases as concentration increases.

For ionic solutes the equation including the correction factor is given as

$$\pi = nRTC\phi \tag{7.19}$$

where n, the van't Hoff factor, is the number of ions produced if a solute molecule completely dissociates.

The spontaneous flow of the solvent across the semi-permeable membrane is in the opposite direction to that required in membrane separations. In reverse osmosis, the aim is to promote the flow of solvent from the more concentrated solution to the less concentrated one. Since osmotic pressure opposes the desired direction of flow, pressure sufficient to nullify the osmotic pressure and overcome the resistance of the membrane is applied on the concentrated solution side (see Figure 7.6c).

The RO membrane is anisotropic with a skin of active membrane of 0.2–0.5 μm thick supported on a porous support film of 50–100 μm thick. The flux J for the solvent is given by:

$$J = A(\Delta P - \Delta\pi) \tag{7.20}$$

where A is a constant depending on the thickness of the membrane and solvent molar concentration, molar volume and mobility. The flux J_s for the solute is not affected by the pressure difference and depends on the Fick's diffusion. It is given as:

$$J_s = k(C_b - C_p) \tag{7.21}$$

where k is a constant depending on the solubility and diffusivity of the solute in the membrane and the membrane thickness.

7.13 DIALYSIS

Dialysis is well known in medical field as hemodialysis. Blood is drawn from a patient and passed through the lumen of the fiber of a hollow fiber unit (artificial kidney) while water containing solutes such as potassium salts, is passed through the shell side. The water contains dissolved salts so that it has the same osmotic pressure as blood to minimize transfer of water. Urea, uric acid, creatinine, phosphates and excess of chloride diffuse from the blood into water thereby purifying the blood of the waste products. Dialysis has potential applications in bioseparations, for example, separation of alcohol from beer. Dialysis involves the separation of solutes by diffusion across the membrane from one liquid phase to another liquid phase, on the basis of molecular size and molecular conformation.

A pressure differential across the membrane is deliberately avoided and the sole driving force is concentration difference across the membrane, the solute diffusing from the more to the less concentrated solution containing the same solvent. The concentration difference between the two solutions should be sufficiently large and the membrane should be thin to reduce the diffusion path. Dialysis membrane has the non-porous characteristics of RO membranes as well as the microporous characteristics of UF membranes. The process is relatively slow compared to UF and RO. It is of interest to note that the solvent passes through the membrane in UF and RO while it is the solute that diffuses through the membrane. Large molecules cannot pass through the membrane.

With concentration difference being the sole driving force, the diffusion coefficients of solutes are inversely proportional to the square root of molecular weight with small molecules diffusing faster than larger ones. The solute flux across the membrane, J_s is given by the equation:

$$J_s = r\Delta C \tag{7.22}$$

where r is the sum of the resistance's offered by the membrane and the two films on the two sides of the membrane (given as $1/r = 1/r_m + 1/r_{f1} + 1/r_{f2}$) and ΔC may be taken as the mean concentration difference of the solute in bulk across the membrane. Solvent flow across the membrane also occurs in dialysis because of osmotic pressure difference ($\Delta\pi$) even in the absence of external pressure gradient (Δp) and this is undesirable as it dilutes the feed and also reduces the solute driving force. Co-current flow is usually used to avoid pressure drop across the membrane.

7.14 ELECTRODIALYSIS

The method was developed for the desalination of brackish water into potable water. Separation of ions occurs due to the imposed potential difference across the ion selective cationic and anionic ion exchange membranes. The electrodialysis unit consists of a stack of compartments formed by alternate cationic and anionic ion exchange membranes (see Figure 7.7). The driving forces is an applied electric field, which induces a current driven ionic flux across the compartments. Electrokinetic

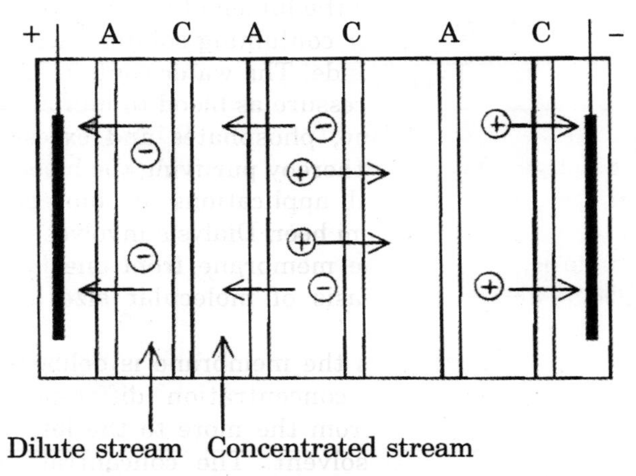

Figure 7.7 Electrodialysis unit.

transport of positively charged species through the cationic membrane and the negatively charged species through the anionic membrane occurs and selective separation of mixtures of ionic species depends on the differing ionic mobilities of the species within the ion exchange membranes. In a continuous flow system, at steady state, the effluents from alternate compartments comprise concentrated and depleted streams of ionic species.

7.15 PERVAPORATION

Pervaporation is a membrane separation technique accompanied by a change of phase of the species transported across the membrane, usually from liquid to vapour. The technique is useful in the separation of liquid components of a mixture such as an azeotrope mixture and is carried out by circulating the liquid to be separated on one side of a coated membrane and vacuum is applied on the other side. Alternatively, an inert carrier gas may be used instead of vacuum which transports the permeate away from the membrane. Separation is brought about due to the differences in the solubility and diffusion of the species in the membrane. As pervaporation is accompanied by a change of phase from liquid to vapour, the heat of vapourization has to be supplied to sustain the process.

7.16 APPLICATIONS OF MEMBRANE SEPARATION PROCESSES

The range of applications of the different membrane separation processes include fruit processing, dairy and pharmaceutical industries based on molecular size of components as summarized in Table 7.2.

TABLE 7.2 Applications of Membrane Separation Processes

Species	*Molecular weight*	*Size* (nm)	*Technique*
Inorganic salts	10–100	0.1–0.2	RO
Simple organic substances (acids, sugars)	100–500	0.4–1.0	RO
Antibiotics	400–1000	0.8–1.2	RO
Biopolymers (proteins, enzymes, polysaccharides)	10^4–10^6	2–10	UF + D
Virus		30–300	UF + D
Colloids		100–1000	UF + MF + D
Bacterial cells		300–10^4	UF + MF + D
Yeasts and fungi		10^3–10^4	MF

Questions

1. Classify the membrane separation processes.
2. What are the advantages of membrane separation processes?
3. Discuss the models applicable to membrane separations.
4. Explain the term rejection coefficient.
5. What are the factors which affect the performance of membranes? How are they minimized?
6. Give a schematic diagram of a membrane separation unit and identify the components.
7. Write a note on the different membrane modules.
8. Discuss the principle and operational aspects of microfiltration.
9. What is ultrafiltration? How is it useful in bioseparations?
10. Explain the principle involved in reverse osmosis.
11. Write notes on dialysis, electrodialysis and pervaporation.

CHAPTER

8

Precipitation

8.1 PRECIPITATION OF PROTEINS

Precipitation involves the conversion of the soluble solutes into insoluble solids, which can be subsequently separated from the liquid by physical methods of separation such as filtration or centrifugation. Precipitation serves primarily as a method for isolation or recovery and concentration of the desired product. Hence the method is often used in the preliminary stages of downstream processing of biological products. The method has been well established for the recovery of bulk proteins. It has inherent advantages in that the methodology and equipment are simple. A wide variety of precipitating agents, which do not denature biological products and bring about precipitation at low concentrations, is available. The method can also be used as a purification step as shown in the case of fractionation of blood plasma proteins. The theoretical principles and applications of precipitation for protein isolation or concentration are discussed in the following sections.

A typical globular protein in aqueous solution exhibits a non-uniform distribution of surface positive and negative charges. The surface also exhibits hydrophilic as well as hydrophobic regions. The solubility of the protein is determined by the interactions between various surface regions with the surrounding water molecules. The stability of protein solutions can be considered on the basis of their colloidal nature. In physiological conditions in cellular fluids of ionic strength of 0.15–0.2 M most of the proteins exist in soluble form. Proteins in their natural condition generally exhibit a net negative charge on the surface and attract positive ions to form the so-called Stern layer of counter ions close to the protein surface. The Stern layer is surrounded by a diffuse Guoy-Chapman layer of mobile counter ions as shown in Figure 8.1.

The stability of the protein-electrolyte colloid is due to the balance between the attractive and repulsive forces between colloidal particles not allowing them to form aggregates. The Stern layer by controlling the effective thickness of the more diffuse outer layer determines the stability

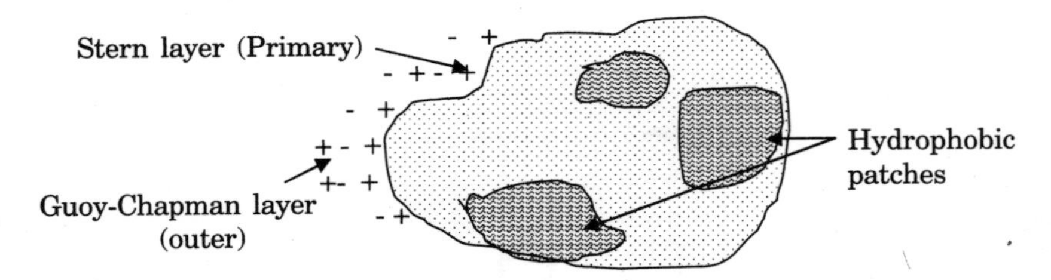

Figure 8.1 The surface of a protein molecule in aqueous environment.

of the colloid. The thickness of the double layer can be reduced by changing the solvent characteristics, particularly the ionic strength and the dielectric constant or by changing the protein surface characteristics thereby bringing about a decrease in the solubility of the protein.

8.2 PROTEIN PRECIPITATION METHODS

The protein precipitation methods may be classified broadly into two groups: (1) Methods in which protein solubility is reduced and precipitation is brought about by altering some physico-chemical property of the solvent such as pH, dielectric constant, ionic strength and water availability. (2) Methods in which protein precipitation is induced by a direct interaction between the protein and a precipitating agent. An overview of protein precipitation methods is shown in the chart (Figure 8.2).

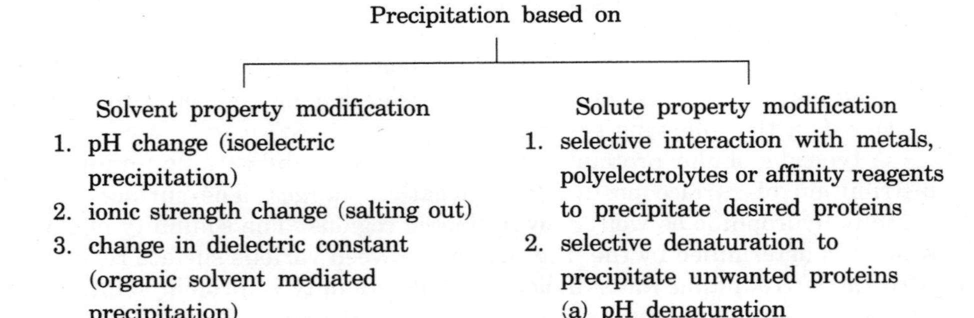

Figure 8.2 An overview of protein precipitation methods.

There are several advantages of protein precipitation as a method of isolation or separation in the initial stages of downstream processing. These include: (i) reduction in the volume of fermentation broth by a factor of 10–50 times (ii) concentration of the desired product (iii) rapid separation and stabilization particularly of labile products (iv) convenience to hold the complex and lengthy downstream processing process at an intermediate stage and diversify into different process steps to achieve product isolation and purification and (v) a less expensive and robust methodology at industrial scale operations to achieve a desired degree of purification. These advantageous features are brought out nicely in the fact that protein precipitation methodology developed in 1940s for the industrial scale fractionation of human blood plasma remains in use even after 60 years. Immunoglobulin G (IgG) and albumin of about 99% purity are obtained from human blood plasma in industrial scale by precipitation methodology alone carried out in five steps (fractional precipitation).

8.2.1 Isoelectric Precipitation

By changing the pH of the protein solution, the ionization of the weak acidic and basic amino acid side chains of a protein is affected resulting in a net zero charge on the protein at a certain pH value called its isoelectric point (pI) or isoelectric pH. When the pH of the protein solution at low ionic strength is adjusted to a value equal to its isoelectric point, solubility of the globular protein decreases drastically over a narrow pH range and it tends to precipitate out. The effect of pH on the solubility of a typical protein such as the soy protein is given in Figure 8.3. The effect of pH change in altering the solubility of a protein is enhanced for proteins of low hydration constant or high surface

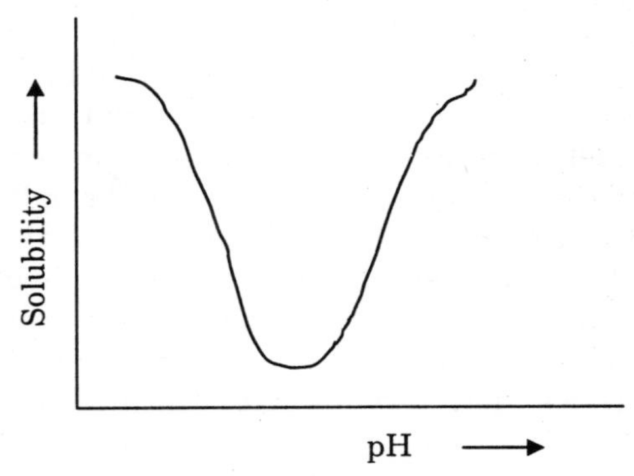

Figure 8.3 Solubility of a globular protein as a function of pH.

hydrophobicity. For example, casein, a protein with high hydrophobicity, precipitates out at its isoelectric point forming large, strong aggregates. On the other hand, gelatin, an exceptionally hydrophilic protein does not precipitate at its pI in low ionic strength or in high dielectric constant media.

The following equation is called Cohn equation:

$$\log S = \beta - kI \tag{8.1}$$

It relates the effect of pH as well as that of neutral salts on the solubility of a protein since the constant β for a particular protein is pH dependent, generally reaching a minimum at or near the isoelectric point of the protein. Sometimes the constant passes through two or more minima, the additional ones being due to the formation of specific salts. It is generally observed that as the neutral salt concentration is increased, the minimum solubility of the protein increases and the pH of lowest solubility decreases due to anion binding. The constant k is called salting-out constant. The ionic strength I, is calculated using the equation

$$I = \frac{1}{2}\sum_{i}^{n} C_i Z_i^2 \tag{8.2}$$

where C is the concentration of the ion and Z is the charge on the ion.

A major advantage of this method is that precipitation is carried out in acid pH range using cheap mineral acids such as hydrochloric, phosphoric, or sulphuric acid. These are acceptable reagents in food products. In addition, it is possible to proceed through to the next step of purification without the necessity of separating the precipitating agent as in the case of salting out of proteins. The main disadvantage of using acids is their potential for denaturing the proteins irreversibly. Further at low pH values the proteins are more sensitive to chaotropic anions which destabilize protein structure.

8.2.2 Protein Precipitation by Addition of Salts

Most enzymes exist in cell fluids as soluble proteins even at high concentrations up to 40%. This solubility is attributed to the combined electrostatic effects consisting of polar interactions with aqueous solvent, ionic interactions with the salts present and to some extent the repulsive electrostatic forces between protein molecules and small aggregates of like charges. The increase in protein solubility with increasing salt concentration in the ionic strength range of zero to about 0.5 M at a given pH and temperature is known as 'salting-in'. The solubility of globulin type proteins at a fixed pH close to its pI varies with salt concentration as shown in Figure 8.4. The curve is referred to as the 'salting-in' curve of a protein.

The salting-in phenomenon can be exploited for purification of proteins in two distinct ways: (i) aggregation and hence precipitation of a protein may be brought about by dilution or dialysis and if the desired protein is present in the aggregate then some extent of purification has been achieved; (ii) isoelectric precipitation of the desired protein (leaving other proteins in solution) may be brought about by adjusting the pH at constant ionic strength thereby enhancing the purity of the product.

Precipitation of proteins by the addition of salt known as 'salting-out' has been shown to be very effective and relatively a cheap method for the fractionation of blood proteins. The method causes little denaturation of

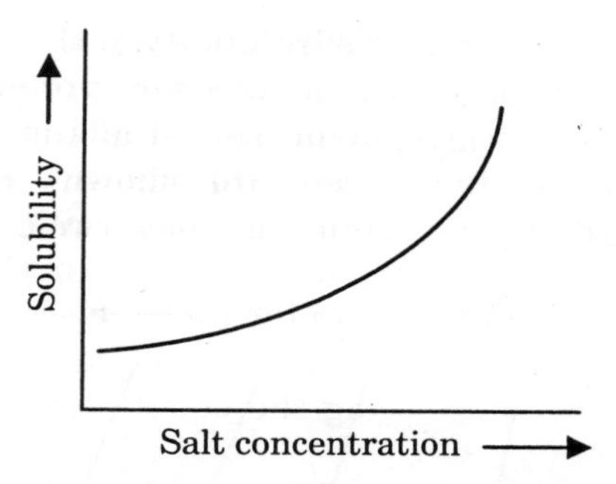

Figure 8.4 Salting-in of proteins at salt concentrations < 0.5 M.

proteins and in fact stabilizes them. The effect of high salt concentration in promoting aggregation and precipitation of proteins has been attributed to the combination of common ion effect, which lowers the solubility and interactions between hydrophobic regions of the protein molecules. The added salts remove water from the hydrated protein, exposing the hydrophobic regions to interact intermolecularly resulting in their aggregation. Hofmeister observed that the charge on the anions as well as cations influenced their ability to bring about precipitation of proteins and formulated a series of ions, referred to as lyotropic series, in order of approximate decreasing effectiveness in causing the precipitation of proteins.

Anions: citrate^{3-} > phosphate^{3-} > tartrate^{2-} > sulphate^{2-} > F^- > IO_3^- > $H_2PO_4^-$ > acetate$^-$ > $B_2O_3^-$ > Cl^- > ClO_3^- > Br^- > NO_3^- > ClO_4^- > I^- > CNS^-

Cations: Th^{4+} > Al^{3+} > H^+ > Ba^{2+} > Sr^{2+} > Ca^{2+} > Mg^{2+} > Cs^+ > Rb^+ > NH_4^+ > K^+ > Na^+ > Li^+

The ions, particularly the anions towards left in the series are better agents for protein precipitation and are known as antichaotropic. On the other hand, the anions towards the right in the series decrease hydrophobic effect by destroying the ordered structure of water and also cause destabilization of protein structure and hence are chaotropic.

Protein precipitation is usually carried out at high ionic strengths and the empirical Cohn equation (Eq. (8.1)) is widely used to predict the solubility of the protein as a function of ionic strength or salt concentration. The protein solubility S is usually expressed in g/l. The constant β in Eq. (8.1) is dependent on the nature of protein, temperature and pH and is independent of the salt used while the salting-out constant k, is independent of pH and temperature but varies with the protein and the salt used. The constants are characteristic of each protein.

According to Cohn equation, pectins, hydrophilic polymers and proteins of similar composition, but increasing molecular weight require less of salt for precipitation. The salting-out curves, actually straight lines, obtained by plotting the log of solubility of the protein as a function of ionic strength or salt concentration, of a series of proteins of increasing molecular weight such as myoglobin, serum albumin, haemoglobin and high molecular weight fibrinogen are shown in Figure 8.5. The asymmetric structure of the protein also has a role in its solubility, the

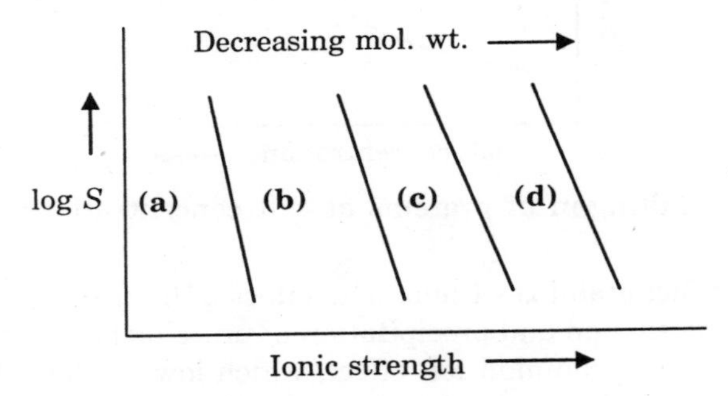

Figure 8.5 Solubility of proteins as a function of ionic strength. (a) fibrinogen (b) haemoglobin (c) serum albumin (d) myoglobin.

greater the asymmetry of the protein, the greater is the degree of precipitation at equivalent salt concentrations. For a particular salt, the salting-out constant k varies only over a two fold range for different proteins, being greatest for large and asymmetric molecules. The solubility of a given protein is also decreased by the presence of other proteins during salting-out.

The salting-out of proteins may be described as a balance between a salting-in process due to electrostatic effects of the salt and a salting-out process due to hydrophobic interactions. Initially at low salt concentrations contribution to electrostatic free energy increases, which favours solubility of protein in the presence of salt. The electrostatic contribution varies non-linearly at low salt concentration and linearly at higher concentrations. Salting-in of proteins is due to this electrostatic effect. The hydrophobic effect predominates at salt concentrations greater than 0.5 M and the protein solubility decreases (salting-out). The relative surface hydrophobicity determines the contact area (Ω) between the protein molecules. At higher salt concentrations the attractive force between the hydrophobic areas is increased due to greater induced dipoles (σ). Simultaneously, the increased electrostatic effect (λ) increases the repulsion between the molecules and the slope of the overall salting-out curve is given by Eq. (8.3).

$$k = \Omega\sigma - \lambda \qquad (8.3)$$

The type of salt used determines the magnitude of σ; this property being governed by the molal surface tension increment of the salt. The Hofmeister series reflects the ability of salts to precipitate proteins on the basis of their molal surface tension increment. The nature of the protein determines the contact area Ω. Generally the magnitude of the product (Ωσ) is much greater than that of λ. Therefore maximum solubility is reached at relatively low salt concentrations, the solubility decreasing rapidly at higher salt concentrations. Figure 8.6 shows the relative

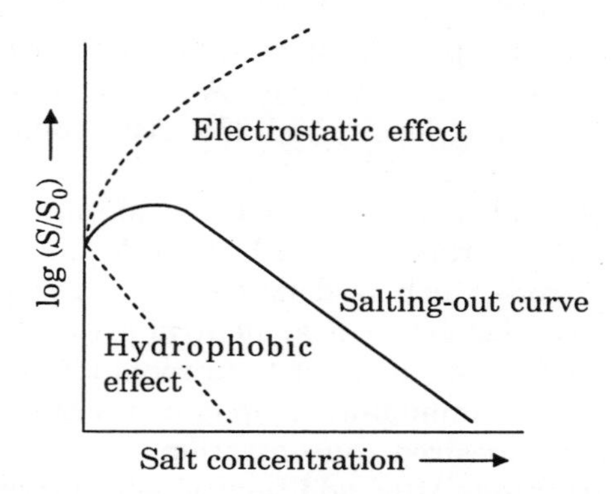

Figure 8.6 Electrostatic and hydrophobic effects on the solubility curve of a protein.

magnitudes of these effects on the slope of the salting-out curve obtained by using the dimensionless form of Cohn equation.

$$\log \frac{S}{S_0} = \beta - kI \quad (8.4)$$

where S_0 is the solubility at zero ionic strength.

In certain cases, the salting-out curve correlates with the relative surface hydrophobicity or the frequency of charged surface groups rather than with the average hydrophobicity of the molecules as shown by proteins such as tetrameric myoglobin (68,000 Da), haemoglobin (68,000 Da) and albumin (67,000 Da), which require different concentrations of ammonium sulphate for precipitation.

Practical aspects. Salts containing polyvalent ions such as sulphates and phosphates have higher values of k than uni-univalent salts, but Ca^{2+} and Mg^{2+} ions depress the value of k. Therefore, ammonium or sodium sulphate and sodium or potassium phosphate have been widely used for salting-out proteins. Sodium sulphate is less soluble at temperatures below 40°C and hence its use is restricted to precipitate thermally stable proteins such as extracellular enzymes. Though salting-out effect of phosphate is greater than that of ammonium sulphate, the latter is more soluble, and its salting-out effect is less dependent on temperature in the range 0–30°C. Another advantage is that a 2–3 M ammonium sulphate suspension of protein precipitate or crystals is often stable for years as the high salt concentration prevents proteolysis and bacterial action. Hence it has become the normal packaging method for commercial enzymes. Further it is cheaper. However it undergoes hydrolysis releasing ammonia at higher pH values, besides being corrosive and difficult to handle and dispose. It also suffers from another disadvantage in that residues of ammonium sulphate remaining in food products can affect the taste even at low concentrations and it is also toxic with respect to clinical use and hence must be removed.

Centrifugation of the protein precipitate will be easier if the densities of the salt and the protein are different. A saturated ammonium sulphate solution is 4.05 M and the density is 1.235 g/cm^3 compared to the protein aggregate density of 1.29 g/cm^3.

Addition of solid salt is always preferred to minimize dilution. EDTA in the concentration range of 10–50 mM is added to prevent contamination by iron which is detrimental to sensitive enzymes. The salting-out is best carried out at near neutral pH (6–7.5) in about 50 mM phosphate buffer, preferably at lower than room temperature to avoid any denaturation. Powdered ammonium sulphate may be added slowly in small increments to reach a desired concentration. Stirring should be efficient without causing frothing. After addition of salt, stirring is continued for about 30 minutes to complete equilibration and then the solution is centrifuged at about 10^5 g/min. i.e., 10,000 g for 10 min or 3000 g for 30 min. The supernatant is removed and the amount of salt to be added to the known volume of supernatant to reach the next per cent saturation is then added in a similar manner. The precipitate is redissolved in an appropriate buffer and if necessary, the solution may be desalted by dialysis or gel filtration.

The amount of ammonium sulphate to be added to reach a required molar concentration M_2 from an initial concentration M_1 has been given by Scopes as:

$$g = \frac{533(M_2 - M_1)}{(4.05 - 0.3\,M_2)} \tag{8.5}$$

where g is the amount (in grams) of ammonium sulphate required. The factor 533 represents the solubility of the salt in grams at 20°C in one litre of water to make a saturated solution. However, 761 g must be added to one litre of water to make a saturated solution as the addition of salt causes an increase in volume to 1.425 litres. The density of the solution is 1761/1425 = 1.235 g/cm^3. Another equation of Scopes gives the quantity of ammonium sulphate to be added for reaching a percentage saturation of S_2 from an initial percentage saturation of S_1 as:

$$g = \frac{533(S_2 - S_1)}{(100 - 0.3\,S_2)} \tag{8.6}$$

Determination of purification factor and per cent yield. A typical small scale trial experiment is necessary to determine the ammonium sulphate required to precipitate out a desired enzyme from a crude extract containing a mixture of proteins and to calculate the purification factor and the per cent yield. After the addition of ammonium sulphate to arrive at a desired concentration or per cent saturation with respect to ammonium sulphate, the amounts of protein and the desired enzyme in the precipitate and supernatant are determined. For example, if the amount of protein precipitated out is 40% of the total protein (including the enzyme) in the extract and the amount of desired enzyme precipitated out at the same per cent saturation of ammonium sulphate is 20% of the total enzyme available in the crude extract, then the purification factor (the ratio of the amount of the desired product to that of the total protein precipitated out under specified conditions) is 0.5. Since the yield of the desired product is only 20% of the total available product, the concentration of ammonium sulphate needs to be enhanced to a higher level. If under this condition the amounts of protein and enzyme precipitated out are 20% and 60% of the available products in the extract, then the purification factor achieved is 3.0 and the enzyme recovery is 60% of the total available quantity. However, under the same conditions if the amounts of protein and enzyme precipitated are 50% and 75%, though the yield of the desired enzyme is higher, the purification factor achieved is only 1.5. Thus fractionation is always a compromise between yield (recovery) and the degree of purification. Depending on the requirements and ready or economic availability of the source material, either yield or purity has to be sacrificed.

It is necessary to mention that salt never precipitates out all the available protein, it only decreases its solubility. Thus about 90% recovery of the total protein may be achieved. Salting-out is more useful for concentrating protein solutions, particularly after gel filtration or ion exchange step. If the starting solutions are very dilute (containing less than 1 mg/ml of protein) concentration by ultrafiltration prior to salting-out will be beneficial.

In practice the Cohn equation has limited ability to predict optimum salt concentration for the separation of proteins from one another. An alternative equation has been suggested by Niktari et al for the salting-out behaviour of yeast alcohol dehydrogenase (ADH) in the presence of other proteins. This is expressed as:

$$E = \frac{1}{1 + (C/\alpha)^{\eta}} \tag{8.7}$$

where E is the fraction of the enzyme remaining in solution, C is the concentration of ammonium sulphate expressed as percentage saturation and α and η are constants. Salting-out of yeast ADH resulted in 80% yield with a purification factor of 3.5 in a two-step fractional precipitation.

8.2.3 Precipitation by the Addition of Organic Solvents

Addition of a water miscible solvent such as ethanol or acetone to an aqueous extract of proteins brings about the precipitation of proteins due to a combination of factors. The primary effect is due to the reduction in water activity and the solvating power of water for a charged hydrophobic protein molecule. This can be quantitatively described in terms of a reduction of dielectric constant of the medium given by Eq. (8.8).

$$\log S = \frac{K}{D^2} + \log S_0 \tag{8.8}$$

where D is the dielectric constant of the reagent-water mixture, K is a constant related to the dielectric constant of the original aqueous medium and S_0 is the extrapolated solubility. The decrease in solubility of proteins leads to aggregation and precipitation mainly due to electrostatic and dipolar interactions between oppositely charged regions of protein molecules. The involvement of electrostatic and dipolar interactions in protein aggregation is supported by the fact that precipitation occurs at lower organic solvent concentration at the isoelectric points of the proteins. Thus the mechanism of aggregation is similar to that in isoelectric precipitation. Hydrophobic interactions seem to play only a minor role in protein precipitation because of their solubilizing effect.

The decrease in the solubility of the protein due to the addition of organic solvent may also be explained on the basis of chemical potential, μ of the solute. At equilibrium the chemical potential of the solid precipitate is the same as that of the species in solution.

$$\begin{aligned} \mu(\text{ppt}) &= \mu(\text{solution}) \\ &= \mu^\circ + RT \ln C \end{aligned} \tag{8.9}$$

where μ° is the standard chemical potential, R is the gas constant, T is the absolute temperature and C is the concentration of the solute. When an organic solvent is added in small quantity to an aqueous solution, the chemical potential of the solute in the mixed medium changes and varies with the solute concentration as shown in Figure 8.7.

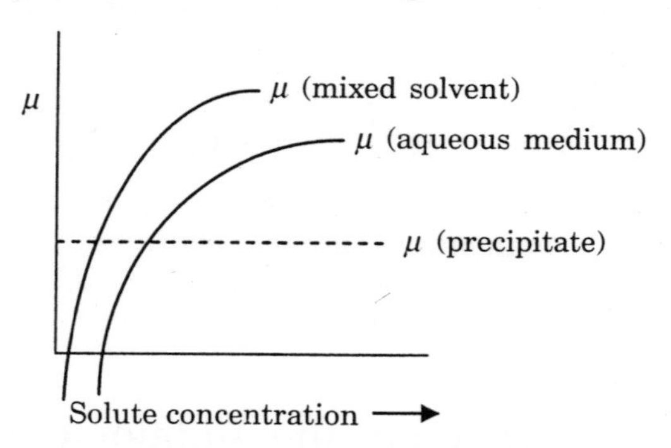

Figure 8.7 Variation of chemical potential in aqueous and mixed organic-aqueous media.

Source: P.A. Belter, E.L. Cussler, and Wei-Shou Hu, *Bioseparations—Downstream Processing for Biotechnology*, John Wiley & Sons, New York, 1988.

The standard state chemical potential of the solute in the mixed medium is raised as a result of which the solution becomes supersaturated at the given temperature and ionic strength. This leads to precipitation of the solute until a new saturated concentration is reached which is lower than that in pure water.

In general, organic solvent mediated precipitation depends on the molecular size of the proteins, the larger the protein molecule, the lower the concentration of organic solvent required to precipitate it. There seems to be an approximately linear relationship (with a negative slope) between log molecular weight of the protein and acetone concentration required for precipitation.

Precipitation of proteins by organic solvents has an advantage in that the variation of dielectric constant together with pH, temperature, ionic strength and protein concentration provides a refined method for protein fractionation. Ethanol and acetone are the most widely used solvents. Other solvents such as methanol, *n*-propanol, iso-propanol, dioxan and 2-methoxyethanol can also be used. Methanol is less effective because of its higher dielectric constant and it is also less specific for precipitating a given protein. Certain organic solvents such as long chain alcohols are more denaturing than short chain ones. Acetone in general has a lesser tendency to cause denaturation than ethanol, possibly because lower concentrations are sufficient to cause precipitation. Temperatures can go to as low as –10°C as in ethanol precipitation of human plasma proteins. At temperatures above +10°C protein denaturation is substantial in organic solvent precipitation.

Addition of a neutral salt to an aqueous-organic solvent mixture produces a salting-in effect and modifies the value of the constant K in Eq. (8.8). As the dielectric constant of the medium decreases, the salting-out effect of the salt decreases rapidly than the salting-in effect. This factor has been taken advantage to bring about fractional precipitation and thereby purification of proteins.

The solubility of metal-protein complexes is very much dependent on the dielectric constant of the medium. This property has been exploited in human plasma protein fractionation where barium and zinc salts have been used to give improved fractionation. The presence of metal ions also reduces the concentration of ethanol required to precipitate the proteins.

The disadvantages of organic solvents include their denaturing effect on proteins, particularly at higher concentrations and at temperatures greater than +10°C. In addition they are inflammable necessitating the use of costly flame proof equipment in large scale operations.

Practical aspects. Precipitation is best carried out at lower temperatures (close to 0°C), at ionic strengths in the range of 0.05–0.2 M and pH at or close to isoelectric point of the desired protein. It is important to take note of the fact that the temperature increases when a cold organic solvent is added to a cold aqueous solution even with efficient stirring and cooling primarily due to hydration of the solvent. The increase in temperature occurs initially as the organic solvent content increases from zero to about 15%, after which there is little heating effect (see Figure 8.8). The initial increase in temperature may denature labile products. However, cellular proteins and enzymes from thermophilic organisms are not denatured by organic solvent precipitation even at elevated temperatures.

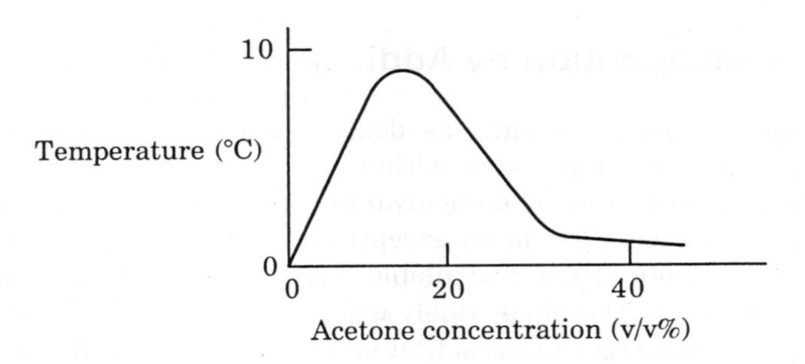

Figure 8.8 Temperature profile as a function of acetone concentration.

Source: R.K. Scopes, *Protein Purification—Principles and Practice*, 3rd ed., Narosa Publishing House, New Delhi, 1994.

Ionic strengths greater than 0.2 M require more solvent to cause precipitation whereas very dilute solutions (ionic strength < 0.05 M) yield a finely divided precipitate which is difficult to filter.

Most proteins have been found to precipitate with acetone in the range of 20 to 50% v/v. It is easy to use additive volumes to express the v/v ratio because the addition of 50 ml of acetone to 50 ml of water results in a total effective volume of about 95 ml only and the balance is lost due to the formation of hydrated solvent complex that occupies smaller volume than the constituent components. Higher molecular weight proteins require less volume of acetone to initiate precipitation. After equilibrating the mixture at the given percentage of solvent at a fixed temperature for about 15 minutes, the precipitate can be centrifuged, preferably in a refrigerated centrifuge. The recovered precipitate can be redissolved in cold buffer. Any residual organic solvent up to about 5% does not affect further fractionation except in hydrophobic interaction chromatography. Higher percentages of organic solvent can be evaporated off at 20–30°C under vacuum or suction.

More solvent can be added to the supernatant obtained from the first precipitation to reach the next desired percentage value. Scopes gives another formula to calculate the volume V (in ml), of the solvent to be added to one litre to take it from an initial concentration of x to a final concentration of y as:

$$V = \frac{1000(y - x)}{100 - y} \tag{8.10}$$

Only proteins of low molecular weight (less than 20 kDa) and proteins whose isoelectric points are far from the precipitation pH will remain in solution. Precipitation of acidic proteins may be aided by the addition of a divalent metal ion such as Mg^{2+}. Alternately, the pH of the medium may be adjusted to near isoelectric point without increasing organic solvent concentration.

8.2.4 Precipitation by Addition of Non-ionic Polymers

Water-soluble polymers such as dextran and polyethyleneglycol (PEG) cause aggregation of proteins without denaturation and can be used at ambient temperatures. The disadvantage of such polymers is their high viscosity. However, PEG is an exception in that solutions up to 20% w/v are not too viscous. PEG is available commercially in a variety of degrees of polymerization, the most widely used are those with molecular weights 6,000 and 20,000 Da. PEG was first used to precipitate the plasma protein fibrinogen, a large highly asymmetric molecule that is just soluble in plasma. Among the other components of plasma, γ-globulins precipitate out first at neutral pH close to their isoelectric points followed by others.

The mechanism of action of these polymers possibly involves reducing the effective amount of water available for the solvation of proteins and excluding the proteins from part of the solution. It has been shown theoretically that the volume of solution accessible to the spherical protein molecules in the presence of rod shaped PEG polymer molecules decreases with increasing polymer concentration. The solubility of protein is proportional to the accessible volume, hence a linear equation, which is similar to that of the linear portion of the Cohn equation has been suggested.

$$\log S = \beta - KC \tag{8.11}$$

where the intercept β is a constant influenced by solution conditions such as pH and ionic strength, while the slope K is also a constant determined by the size of the protein and the type of PEG and C is concentration of PEG. Thus a rapid decrease in the solubility (large K) for high molecular weight proteins can be expected in the presence of large molecular weight polymers. An approximately linear decrease in the solubility of the protein (as $\log S$) with increasing polymer concentration (w/v%) is generally observed as predicted by Eq. (8.11). Non-linearity has been observed both at low and high protein concentrations.

At high concentrations Foster equation (Eq. (8.12)) may be used to analyze PEG mediated precipitation of proteins as in the case of yeast intracellular proteins.

$$\ln S + fS = \beta - KC \tag{8.12}$$

where S is the protein solubility, f is protein self-interaction coefficient which is higher at higher protein concentrations (> 10 mg/ml) and at $pH \neq pI$.

In general, higher concentrations of PEG are required to precipitate proteins of low molecular weight and vice versa. Precipitation generally occurs by the addition of concentrated aqueous solution of PEG at concentrations less than 20%. As with organic solvents, proteins become more soluble in PEG solutions as the pH moves away from the pI. The presence of divalent metal ions also influence the solubility of various proteins in PEG solutions.

Proteins precipitated by non-ionic polymers can be processed by ion-exchange resins directly without any intermediate step to remove the precipitating agent as is necessary in the case of salting-out of proteins. PEG is not retained by ion exchange resins. PEG of 4000 molecular weight is accepted as a safe precipitant for human plasma. A residual low level of PEG is not detrimental to many other purification procedures such as affinity chromatography and gel filtration. Higher concentrations of PEG, however, can adversely affect the performance of exclusion media such as Sephadex G-100. If necessary, PEG in protein solutions can be removed as a separate phase by addition of 0.4 M phosphate while the protein is retained in the aqueous phase or by the addition of ethanol in which PEG is more soluble whereas the protein is not soluble.

8.2.5 Precipitation by Ionic Polyelectrolytes

Ionic polyelectrolytes act similar to flocculating agents and are gainfully employed in industry for the recovery of enzymes and food proteins. Electrostatic forces are primarily responsible for protein precipitation. The advantages of using polyelectrolytes include the requirement of very low concentrations in the range of 0.05 to 0.1% (w/v) to cause precipitation, their low price and absence of waste disposal problems. Examples of polyelectrolytes include acidic polysaccharides such as alginate, pectate, carboxymethylcellulose and carageenan; anionic polyelectrolytes such as polyacrylic acid (PA) and polymethacrylic acid (PMA); cationic polyelectrolytes such as polyethyleneimine (PEI); and polystyrene based quaternary ammonium salts (PSQA). The effect of pH on whey protein precipitation brought about by addition of polyelectrolytes is shown in Figure 8.9. Basic proteins such as lysozyme, cytochrome C, prolamines and trypsin are precipitated by polyacrylic acid which can be removed by precipitating as calcium salt at pH 6. The recovery of the enzymes is over 90%. Polyethyleneimine has been found to be useful in the purification of enzymes and the polyelectrolyte can be removed by cation exchange or by complexing with anions. Polyphosphates have been used for the precipitation of plasma proteins other than gamma globulins at pH 4 to 5.

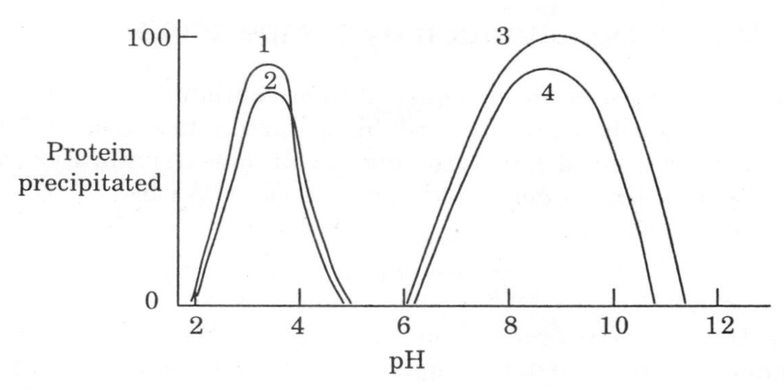

Figure 8.9 Whey protein precipitation by the addition of polyelectrolytes. 1. PA 2. PMA 3. PSQA 4. PEI.

8.2.6 Precipitation by Metal Ions

Polyvalent metal ions cause protein precipitation. Based on the mode of action they may be classified into three groups characterized by: (i) ions such as divalent manganese, iron, cobalt, nickel, copper, zinc and cadmium bind strongly to carboxylic acids and to nitrogenous groups such as amines and heterocyclic groups; (ii) ions such as divalent calcium, barium, magnesium and lead bind to carboxylic acids but not significantly to nitrogenous ligands; and (iii) ions such as monovalent silver, mercuric mercury and lead bind strongly to sulphydryl groups. An advantage of metal ion precipitation is that metal ions have greater precipitating power with respect to proteins even in dilute solutions and in addition they can be easily removed subsequently by chelating agents or cation exchange resins. Calcium, barium and zinc salts have been used to modulate ethanol precipitation of human plasma proteins.

8.3 SELECTIVE DENATURATION OF UNWANTED PROTEINS

Many proteins are stable to extremes of pH or temperature conditions. When such proteins or enzymes are the desired ones they can be purified by retaining them in solution while the unwanted proteins can be destabilized and precipitated out by selective denaturation. Denaturation of a protein is due to the destruction of the tertiary/quaternary structure resulting in the formation of a random coiled polypeptide chain. Such denatured polypeptide chains aggregate easily and have relatively low solubility. Selective denaturation of unwanted proteins and their precipitation can be brought about by (i) higher temperature (ii) extremes of pH and (iii) addition of organic solvents at higher temperature in the range of 25–30°C.

8.3.1 Selective Denaturation by Temperature

Increase in temperature of an aqueous solution containing a mixture of proteins can selectively denature and hence precipitate some of the proteins leaving the other desired proteins in solution. Denaturation of a protein by heat is a first order kinetic process described by:

$$\frac{d[P]}{dt} = -k[P] \tag{8.13}$$

where $[P]$ is the protein concentration in solution and k is the first order rate constant with an Arrhenius temperature dependence as given by:

$$k = k_0\, e^{-E/RT} \quad \text{or} \quad \frac{d[\ln k]}{dT} = \frac{E}{RT^2} \tag{8.14}$$

where k_0 is a constant, E is the activation energy for denaturation related to the enthalpy of activation ($\Delta H^{\#}$) and temperature T as $E = \Delta H^{\#} + RT$. The energy of activation for thermal denaturation of proteins is quite high, in the range of 200–400 kJ/mol compared to energy of activation for chemical and enzyme catalyzed reactions in the range of 40–70 kJ/mol. This may be attributed to the large positive entropy change when a protein unfolds due to denaturation. As E varies exponentially with temperature, even small changes in temperature have profound influence on the extent of denaturation of proteins. The extent of denaturation of a protein as a function of temperature is shown in Figure 8.10.

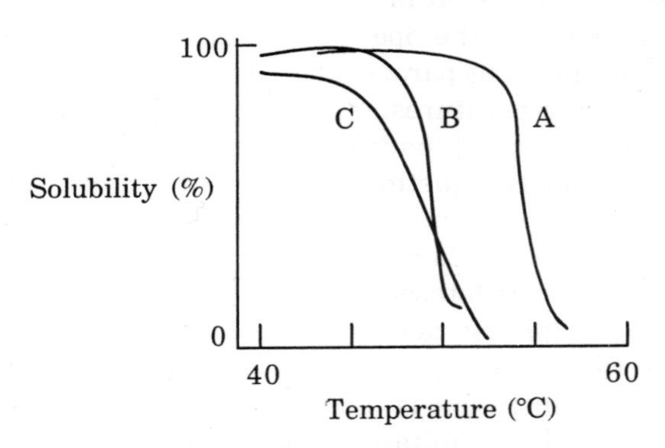

Figure 8.10 Thermal denaturation curve of proteins.

The thermal denaturation curve of the protein is obtained by incubating the protein for about 10 minutes at each of the selected temperatures and measuring the amount of protein remaining in solution after centrifuging off the precipitated protein. The energy of activation for denaturation of the protein is calculated from the slope of the linear portion of the curve. From a knowledge of the thermal denaturation curves of individual proteins it is possible to choose a temperature (e.g.

50°C) at which the desired protein (A) in a given mixture can be retained in solution allowing the unwanted ones (B and C) to undergo denaturation and precipitate out (see Figure 8.10).

8.3.2 Selective pH Denaturation

Extremes of pH cause protein denaturation. This is because the sensitive areas of the protein molecule acquire more like charges thereby causing opening-up of the tertiary/quaternary structure of the protein to a random coil due to electrostatic repulsions. It is also possible that the sensitive areas lose charges which are responsible for holding the protein structure through attractive forces and thereby denature the protein. The most stable pH of a protein need not necessarily be its isoelectric point. For example, many bacterial and plant proteins have isoelectric points in the pH range of 4–5 but are unstable at that pH. They exist in the pH range of 6–7 in the living cell.

Enzymes, generally, become less stable at pH values farther away from their physiological operating pH. For example, papain is very stable and active at pH 1–2 but denatures quickly at pH 7 or higher. It must also be pointed that the optimum pH for stability and optimum pH for activity may not be the same, since the latter determined experimentally in vitro is not relevant physiologically.

Changes in the pH brought about by the addition of acid or alkali involves the protonation or deprotonation reaction which is an extremely fast step. The rate determining step in pH as well as heat denaturation of proteins is the same, namely, the opening-up of the protein molecule. Thus, temperature is an important parameter and pH denaturation should be carried out at specified temperatures. Lower temperatures in the range of 0–10°C are better than 50–60°C, particularly at the extremes of pH.

Protein denaturation by pH adjustment is preferably done by addition of tris and acetic acid in the pH range 4.5–8.5. Lactic acid is suitable up to pH 3.5 and for pH >8, diethanolamine (up to pH 9), sodium carbonate (up to pH 10.5) and sodium or potassium hydroxide (above pH 11) may be used. Phosphoric or sulphuric acid may be used for the pH range 2–3. The denatured proteins do not precipitate completely in the extremes of pH even if the salt concentration is moderate. But on readjusting the pH to neutrality, they will precipitate out. Hence it is advisable to incubate for about 30 minutes after neutralization to ensure complete precipitation. The desired proteins which remain in solution are recovered by centrifuging off the precipitated denatured proteins.

8.3.3 Selective Denaturation by Organic Solvents

Selective denaturation of proteins by the addition of organic solvents is based on the differential sensitivity of proteins towards treatment with organic solvents. Precipitation of the unwanted proteins in a mixture may be effected by the addition of water miscible organic solvents such as alcohols usually at higher temperatures in the range of 25–30°C or even higher. In contrast, the same organic solvents precipitate the desired proteins with little denaturation if the temperatures are close to 0°C or less. For example, the fact that rabbit muscle creatine kinase is stable to 60% v/v ethanol at 25°C in alkaline conditions, was used in the original purification of this enzyme. Similarly yeast alcohol dehydrogenase can be purified by this method because it is stable to ethanol concentration up to 33% v/v at 25°C.

The method can be adopted after ascertaining the sensitivity of the proteins to different organic solvents in preliminary trials. For example, the stability of yeast glyceraldehyde phosphate dehydrogenase towards different alcohols at varying concentrations was determined by incubating the enzyme (2 mg/ml) at 30°C for 30 minutes after which any precipitate formed due to denaturation was centrifuged off and the activity of the enzyme in the solution determined. The experiments indicated that the longer the aliphatic chain, the more denaturing the alcohol was and the percentage of *n*-alcohol required to cause 50% denaturation under the experimental conditions decreased by a factor of exactly 2 as the aliphatic chain length increased by one methylene group. Thus the v/v percentages of methanol, ethanol, *n*-propanol, *n*-butanol and *n*-pentanol required to cause 50% denaturation were 34, 17, 8.5, 4.2 and 2.2 respectively. It is clear that the prime cause responsible for denaturation is the increase in hydrophobicity as the chain length increases (see Figure 8.11). Branched chain alcohols were found to be less denaturing and also more water miscible. Acetone was found to behave very similar to ethanol.

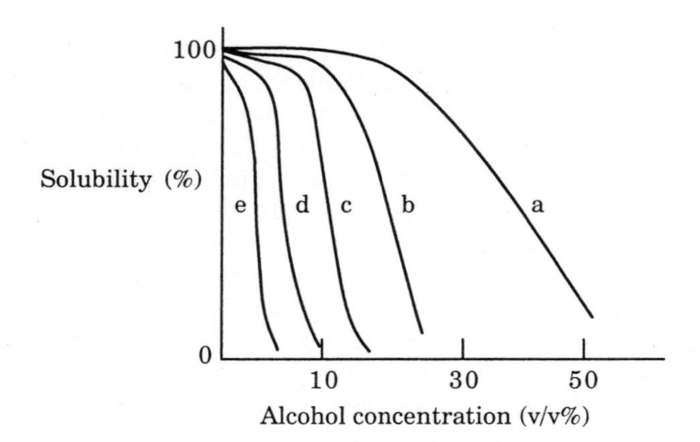

Figure 8.11 Protein solubility as a function of concentration of different alcohols (a) methanol (b) ethanol (c) 1-propanol (d) 1-butanol and (e) 1-pentanol.

Source: R.K. Scopes, *Protein Purification—Principles and Practice*, 3rd ed., Narosa Publishing House, New Delhi, 1994.

The extent of denaturation of proteins by organic solvents is influenced by pH and temperature and hence it is necessary to maintain both these parameters at the required level in effecting the precipitation of unwanted proteins.

An interesting example of organic solvent denaturation using a water immiscible solvent is the selective denaturation of haemoglobin by chloroform. The method is used to recover and purify erythrocyte enzymes in the presence of a large excess of haemoglobin. The haemoglobin is precipitated out by shaking the red blood cell extract with ethanol-chloroform mixture, retaining the desired enzymes in solution.

8.4 LARGE SCALE PRECIPITATION (PROCESS SCALE-UP)

In bioprocesses scale-up means simply a scale which is about ten or more times the laboratory scale and need not necessarily involve tons or even kilogram quantities. Large scale precipitations are subject to the same equilibrium conditions as their small scale parallels, but the kinetics of large scale processes can change. The aim in scaling-up is to maintain the kinetics same in both small scale as well as in large scale operations. The kinetics of large scale precipitations can be maintained the same as in small scale level by using appropriate power per volume. The steps of relevance in large scale precipitations which influence the kinetics include: (i) initial mixing of the feed solution with precipitating agent such as an organic solvent or salt; (ii) nucleation; (iii) growth of precipitate particles under diffusion controlled and flow influenced regimes; (iv) flocculation and (v) centrifugation.

Mixing. Mixing even at high speeds of agitation will not produce a homogeneous solution instantaneously at the molecular level. The time, t, required to produce such homogeneity, depends on the diffusion coefficient of the solute, D, the average size of the turbulent eddies caused by mixing, l, which in turn depends on the solution density, ρ, its kinematic viscosity, ν and power per volume of stirring, P/V as given by

$$t = \frac{1}{4D}\left(\frac{\rho v^3}{P/V}\right)^{1/2} \tag{8.15}$$

where $l = \left(\frac{\rho v^3}{P/V}\right)^{1/4}$ and hence $t = l^2/4\text{D}$

Nucleation. This is the part of the process where small particles appear and begin to grow. In colloidal systems as in most of the biologicals the process is instantaneous.

Growth of precipitate particles (aggregation). The growth of the particles occurs due to their collision brought about by their random motion promoted by thermal energy of molecules. Removal of any hydration or electrical barrier to collision will allow association of the protein molecules leading to aggregation. Initially the rate of molecular association is controlled by the diffusivity of the colliding species. As the particles grow to a bigger size, further growth depends on collision promoted by fluid motion. Thus the growth of precipitate particles can be studied under diffusion controlled conditions called perikinetic growth and flow induced conditions called orthokinetic growth.

Perikinetic growth. In a mono-sized dispersion, the initial rate of decrease of molar concentration C or particle number N, due to the growth of precipitate particles under diffusion controlled conditions follows a second order process according to the theory of Smoluchowski.

$$-\frac{dC}{dt} = kC^2 \quad \text{or} \quad -\frac{dN}{dt} = k'N^2 \tag{8.16}$$

where k', the rate constant is determined by the diffusivity, D and diameter, d, of the particles as $k' = 8\ \pi Dd$. The average diffusion coefficient of all the particles, D, drops as the particles become larger. However, its decrease is inversely proportional to the particle diameter, d and so the product Dd and hence the rate constant k' remain nearly constant. The equations apply to limiting particle size defined by the fluid motion in the range of 0.1–10 μm for high and low shear fields respectively. Any electrical barrier around the particles will reduce the rate of association and this is taken into account by Fuch's modification of Smoluchowski's theory as:

$$-\frac{dN}{dt} = \left(\frac{k'}{w}\right)N^2 \tag{8.17}$$

where w, the stability ratio is given by $d \int_0^\infty \left[\exp \frac{\{\varphi(h)/kT\}}{(h+d)^2} \right] dh$

where h is the particle separation distance, k is Boltzmann constant and $\varphi(h)$ is potential energy interaction between two particles. The value of the stability ratio cannot be evaluated readily as the energy term varies as particle separation decreases. Integrating Eq. (8.16) gives:

$$\frac{1}{C} = \frac{1}{C_0} + k't \tag{8.18}$$

where C_0 is the initial solute concentration. However, it is often difficult to measure C. To overcome this difficulty in solving the Eq. (8.18), the concept of precipitate particle average molecular weight is used. The particle average molecular weight increases linearly and rapidly with time after an initial lag. Using the mass balance relationship,

$$CM(t) = C_0\, M(0) \tag{8.19}$$

where $M(0)$ and $M(t)$ are the average molecular weights of the solute at time $t = 0$ and at t, respectively, the growth kinetics of Eq. (8.18) may be rewritten as:

$$M(t) = M(0)(1 + C_0 k't) \tag{8.20}$$

The theory predicts that the linear increase in $M(t)$ in a stagnant fluid of Brownian particles is a diffusion driven or perikinetic growth in which the diminution of solute concentration C follows a second order process. Experimental evidence for such a relationship has been obtained for calcium ion induced aggregation of casein as shown in Figure 8.12.

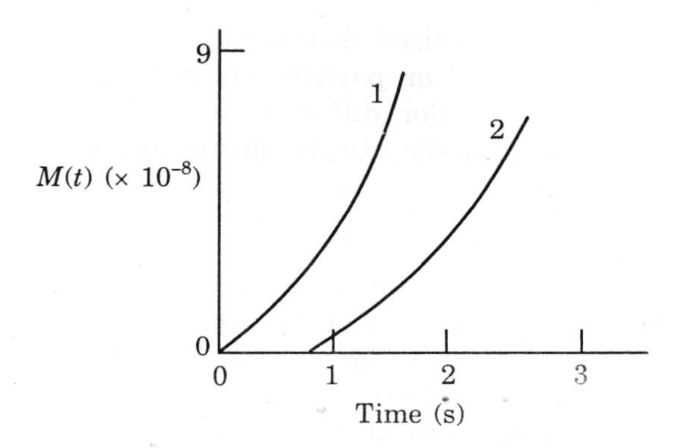

Figure 8.12 Calcium chloride (0.008 M) induced precipitation of casein of (1) 1.0 M and (2) 0.5 M.

Source: J.E. Bailey and D.F. Ollis, *Biochemical Engineering Fundamentals*, 2nd ed., McGraw Hill, New York, 1986.

In the initial stages of the particle growth, calcium ions bind to the casein molecules and through their free valences provide multifunctional points of aggregation. The rate of collision is limited by the number of free valences till a stage where further growth of the precipitate particles is diffusion controlled. The theory does not take into consideration any residual repulsion or hydration barrier to collisions and all collisions are considered to be effective. This relationship was followed up to the formation of casein particles of mean molecular weight of 10^9 or particle size of 0.1 μm. The relation between particle diameter, d, and mean molecular weight may be expressed as:

$$M = 0.105 \times 10^{12} d^3 \tag{8.21}$$

and the corresponding relationship for globular proteins may be expressed as:

$$M = 0.235 \times 10^{12} d^3 \tag{8.22}$$

Orthokinetic aggregation. This is also known as flow induced growth. Larger particles of size greater than about 1 μm grow further by colliding with each other brought about by fluid motion due to mixing. The collisions can also promote breaking up of aggregates due to shear forces caused by mixing and hence the mixing should be slow. The growth of particles under such flow induced conditions follows second order kinetics represented by Eq. (8.16), but the magnitude of the rate constant k' is different and equal to $2/3\ \alpha\, d^3\, G$, where G is the mean shear rate and α is 'sticking coefficient' or collision effectiveness factor introduced to account for the fact that all collisions do not produce growth due to aggregation. The second order kinetic equation can be simplified by assuming a constant volume fraction of particles, $\varphi_v = 1/6\ \pi d^3 C$ (or $1/6\ \pi d^3 N$) and rewritten as:

$$-\frac{dC}{dt} = \frac{4}{\pi}\alpha\varphi_v GC \tag{8.23}$$

which indicates that the change in solute concentration in the flow induced growth regime follows first order kinetics.

Flocculation. This is also known as aging and is the final step in the agglomeration of particles to form larger flocs. No theory can predict the growth of flocs and hence experimental study is necessary. However, using Smoluchowski's orthokinetic growth theory it is predicted that slower rate of flocculation may follow a first order removal of individual particles in to larger flocs. This is achieved by slow mixing. The aging parameter is defined as Gt (where G is the mean shear rate and t is time). Shear forces allow compaction of flocs. Fresh precipitates subjected to an exposure of Gt of at least 10^5 compact the flocs for soya protein as shown in Figure 8.13. Greater shear forces destroy the flocs.

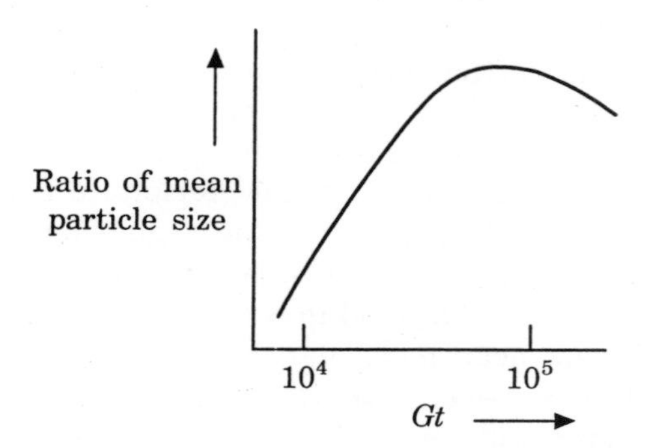

Figure 8.13 Aging of soya protein precipitate. (Protein concentration 30 kg/m^3; particle dia 15–50 μm.)

Source: J.E. Bailey and D.F. Ollis, *Biochemical Engineering Fundamentals*, 2nd ed., McGraw Hill, New York, 1986.

The optimization of aging time and shear rate is necessary to allow the growth of the flocs to a minimum size which may be determined as shown in Figure 8.14.

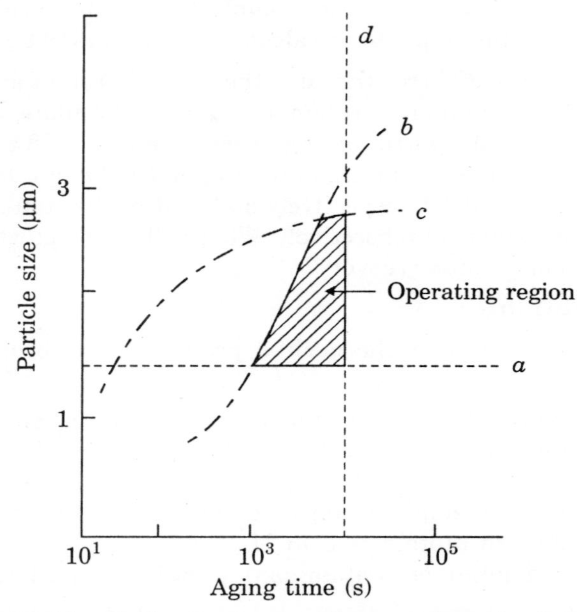

Figure 8.14 Optimum range of parameters for large scale precipitation of proteins on the basis of floc size versus aging time. (a) minimum size for centrifugal recovery (b) optimum strength parameter ($Gt = 10^5$) (c) minimum stirring speed required for uniform mixing and (d) maximum acceptable aging time.

Source: J.E. Bailey and D.F. Ollis, *Biochemical Engineering Fundamentals*, 2nd ed., McGraw Hill, New York, 1986.

Recovery of protein precipitate. Centrifugation is the usual unit operation used to collect the protein precipitate.

Exercises

1. The solubility of a protein is 15 g/litre at ammonium sulphate concentration of 2.2 M and 0.25 g/litre at 3.0 M. Calculate the solubility of the protein at 3.8 M of the salt.

   Solution

   Cohn equation is used to calculate the constants β and k.

   Ionic strength of ammonium sulphate solution is 3 × molar concentration.

$$\log 15 = \beta - k(3 \times 2.2)$$

$$\log 0.25 = \beta - k(3 \times 3)$$

   The values of β and k are 6.058 and 0.74 respectively.

Using these values the solubility of the protein at 3.5 M ammonium sulphate is calculated to be **0.0194** g/litre.

2. A broth of 80 litres contains the desired protein at 12.8 g/l as well as a contaminant protein at 1.8 g/l. Calculate the ammonium sulphate concentration required to recover 98% of the desired protein if the precipitation constants β and k of the desired protein are 9.33 and 1.1 respectively and that of the contaminant protein are 8.8 and 0.95 respectively. What will be the purity of the desired protein at 98% recovery?

Solution

Total amount of the desired protein in the broth = 80 × 12.8
= 1024 g

The amount of protein remaining in solution after 98% recovery by precipitation = 1024 – 1003.5 = 20.5 g

log 20.5 = 9.33 – 1.1 (3 × ammonium sulphate molarity)

Ammonium sulphate concentration required to precipitate 98% of the desired protein = 2.43 M

Total amount of contaminant protein in the broth = 80 × 1.8
= 144 g

At 2.43 M ammonium sulphate concentration 32.36 g of the contaminant protein will be precipitated.

The total amount of the precipitate of two proteins = 1056.36 g

The percentage purity of the desired protein precipitate = **96.94**

Questions

1. Give an outline of the protein precipitation methods.
2. What is isoelectric precipitation? What are its advantages?
3. Explain the terms 'salting-in' and 'salting-out' of proteins.
4. Discuss the theoretical principles and practice of salting-out of proteins by ammonium sulphate.
5. How does the addition of an organic solvent to an aqueous solution of proteins bring about the precipitation of a desired protein?
6. Write a note on the use of non-ionic polymers in protein precipitation.
7. What is selective denaturation of proteins? How is it carried out?
8. Discuss the principle and mechanism of precipitation of proteins by selective thermal denaturation.
9. How does the addition of polyelectrolytes affect the solubility of proteins?
10. Give an account of the steps involved in large scale precipitation of proteins.

CHAPTER

9

Chromatography: Principles and Practice

9.1 CHROMATOGRPHY—A SEPARATION TECHNIQUE

Chromatography refers to a group of closely related separation techniques, which are quite useful in the analysis of chemical substances. The techniques find use in the separation, purification and identification of compounds before quantitative analysis is taken up.

Selective distribution of the components of a mixture between two immiscible phases in intimate contact with each other forms the basis of separation in any chromatographic technique. One of these phases, called the stationary phase is a solid or an immobilized liquid (i.e., a liquid coated on a finely divided inert solid). The other phase is called the mobile phase (eluent or carrier gas depending on whether it is a liquid or gas) and percolates through the stationary phase. The sample is usually dissolved in the mobile phase.

9.2 CLASSIFICATION OF CHROMATOGRAPHIC TECHNIQUES

Chromatographic methods may be classified in three different ways. The first classification is based on the physical configuration and the method by which the stationary and the mobile phases are brought into contact with each other and includes two types, namely, (i) column chromatography and (ii) planar chromatography, as summarized in Figure 9.1.

In column chromatography the stationary phase is packed in a narrow bore tube or column through which the mobile phase (a liquid or gas) flows by gravity or under pressure at a constant flow rate. Many types of chromatographic techniques such as adsorption chromatography, gas chromatography, ion-exchange chromatography, HPLC and gel

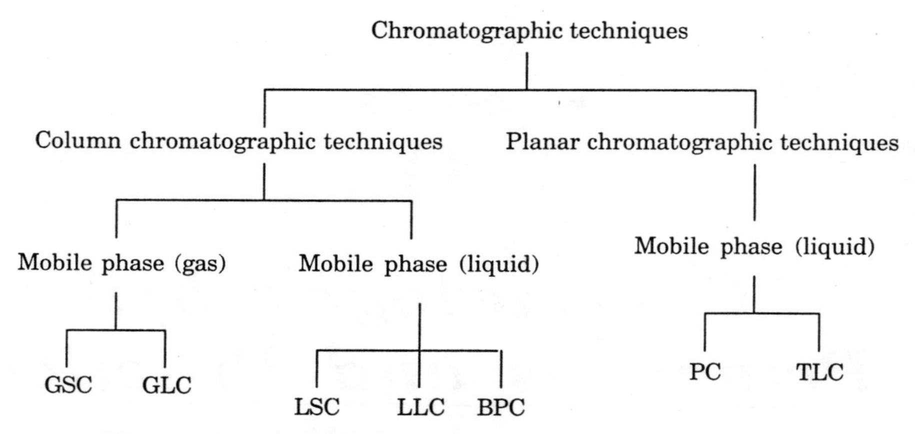

Figure 9.1 Classification of chromatographic techniques. **BPC**—bonded phase chromatography; **GLC**—gas liquid chromatography; **GSC**—gas solid chromatography; **LSC**—liquid solid chromatography; **LLC**—liquid-liquid chromatography; **PC**—paper chromatography; **TLC**—thin layer chromatography.

permeation chromatography come under this classification. In planar chromatography, the stationary phase is supported on a flat plate or in the interstices of a filter paper. The mobile phase moves through the stationary phase by capillary suction or by gravity. Paper chromatography and thin layer chromatography belong to this category.

The second method of classification is based on the nature of mobile phase. Accordingly three general chromatographic techniques can be identified as (i) liquid chromatography (ii) gas chromatography and (iii) supercritical fluid chromatography.

The third classification is more fundamental and is based on the type of interaction involved and the mechanism of separation. When the stationary phase is a solid, the physical phenomenon responsible for the separation of the components irrespective of whether the mobile phase is a gas or a liquid, is the difference in the strength of the adsorption of the components on the solid stationary phase. Such experimental techniques come under the general category of adsorption chromatography. Gas-solid chromatography, thin layer chromatography and liquid-solid adsorption chromatography including the adsorption mode of HPLC are examples of adsorption chromatography. When the stationary phase is an immobilized liquid, separation is achieved by partitioning the components between the stationary and the mobile phases and hence the process is known as partition chromatography. Partition reflects the relative attraction or repulsion that molecules of the competing phases show for the solute components. These attractive or repulsive forces may be polar in nature, arising from permanent or induced electric fields associated with both solute and solvent molecules or due to London dispersion forces, which depend on the relative masses of the solute and solvent molecules. Examples of partition chromatography are gas-liquid chromatography, paper chromatography and liquid-liquid partition chromatography including the normal phase mode of HPLC. In addition to adsorption and partition, other mechanisms of separation involved in chromatographic techniques include ion-exchange, size exclusion, bioaffinity, pseudo-affinity and hydrophobic interaction. A summary of the classification methods is listed in Table 9.1.

Table 9.1 Classification of Chromatographic Techniques

Stationary phase	*Mechanism of separation*	*Specific technique*
Liquid chromatography (mobile phase is a liquid)		
Solid	Adsorption	LSC, TLC
Adsorbed liquid	Partition	PC, LLC, normal phase HPLC
Organic bonded phase	Partition	BPC
	Bioaffinity	AC
	Hydrophobic interaction	HIC, RPC, reversed phase HPLC
Ion-exchange resin	Ion-exchange	IEC
Liquid in the pores of a polymeric bead	Size-exclusion	GPC, Gel filtration
Gas Chromatography (mobile phase is a gas)		
Solid	Adsorption	GSC
Adsorbed liquid or organic bonded phase	Partition	GC or GLC
Super critical fluid chromatography (mobile phase is a supercritical fluid)		
Organic bonded phase	Partition	SCFC

AC—affinity chromatography; **GC**—gas chromatography; **GPC**—gel permeation chromatography; **GSC**—gas solid chromatography; **HIC**—hydrophobic interaction chromatography; **HPLC**—high performance liquid chromatography; **IEC**—ion-exchange chromatography; **LLC**—liquid-liquid chromatography; **RPC**—reverse phase chromatography; **SCFC**—supercritical fluid chromatography; **TLC**—thin layer chromatography.

9.3 GENERAL DESCRIPTION OF COLUMN CHROMATOGRAPHY

A simple experimental set-up of column chromatography is shown in

Figure 9.2. The main components are (1) solvent or eluent reservoir (2) pump (e.g. peristaltic pump) (3) sample injection port (4) column containing the packed stationary phase (5) a suitable detector (e.g. spectral or fluorescence detector) (6) a data processor with display or recorder provision and (7) fraction collector. The eluent is delivered by the pump, from the reservoir to the head of the column, at a constant flow rate.

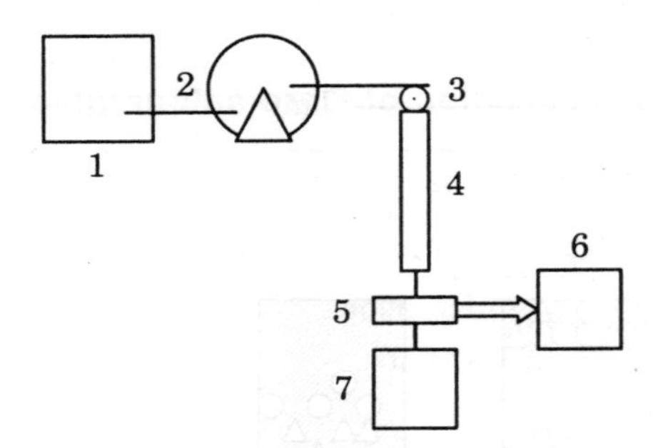

Figure 9.2 A block diagram of a chromatographic set up.

Chromatography may be performed by any of the three methods (i) elution analysis, (ii) frontal analysis or (iii) displacement analysis. The aims of the techniques and methodologies are different.

In *elution analysis* a small volume of the sample of a mixture of components is applied to the column followed by elution (percolation) with a chosen mobile phase. The primary aim of elution analysis is to separate the individual components of the mixture for identification i.e., qualitative analysis and later possibly quantitative estimation of the desired component.

A general description of a classical column chromatographic separation of a two-component mixture A and B is as follows. In elution analysis, the sample solution in the mobile phase or a compatible solvent is introduced as a pulse of about 3–5% of the column volume at the head of the column packed with the stationary phase. Immediately the components of the sample solution distribute themselves between the two phases. Once the adsorption of the sample is complete, the mobile phase is introduced continuously, at a constant flow rate, at the head of the column. The addition of the mobile phase results in the redistribution of the solute components and their simultaneous migration down the column along with the mobile phase. The solute components are exposed to fresh surfaces of the stationary phase leading once again to the partitioning of the components between the two phases. A series of such distributions between the mobile and the stationary phases occurs as the solute components move down the column leading to different migration rates of the solutes A and B depending on the fraction of time a particular solute remains in the mobile phase. The different migration rates lead to the separation of the components A and B into separate bands, which elute out of the column at different times. The schematic representation of the separation of a two-component mixture is shown in Figure 9.3 and the on-line chromatogram in Figure 9.4. The chromatogram is actually a plot or graphical presentation of the variation of the concentration of the solute components as a function of time or volume of eluent as they come out of the column. The concentration profile is obtained by including a suitable detector at the exit end of the column.

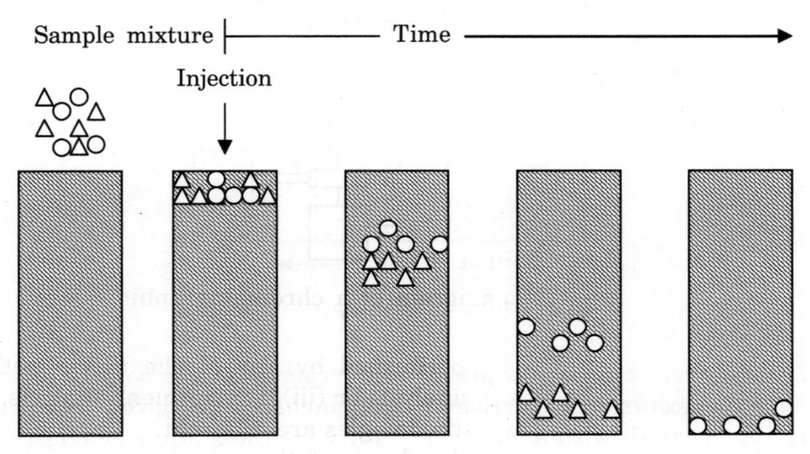

Figure 9.3 A schematic representation of chromatographic separation of a two-component sample mixture.

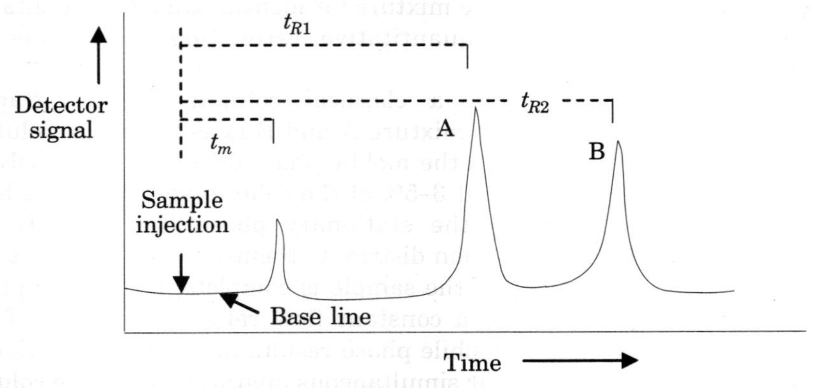

Figure 9.4 The on-line chromatogram of the two component mixture.

The chromatogram shows peaks corresponding to the elution of the different solute components of the mixture, indicated as A and B. The small peak is due to any totally unretained component of the mixture. The detector signal is proportional to the amount of the individual solute and is calculated from the area of the peak for the specific solute. The significance of the terms t_m, t_{R1} and t_{R2} will be discussed shortly.

In *frontal analysis* or *frontal development* the sample of the mixture of components is allowed to percolate through the column continuously with the specific aim to concentrate the component on the stationary phase for recovery at a later stage in a concentrated form. The frontal development of a three component mixture is shown in Figure 9.5. The weakly adsorbed component A is discharged from the column first followed by a mixture of A and B and finally by a mixture of A, B and C.

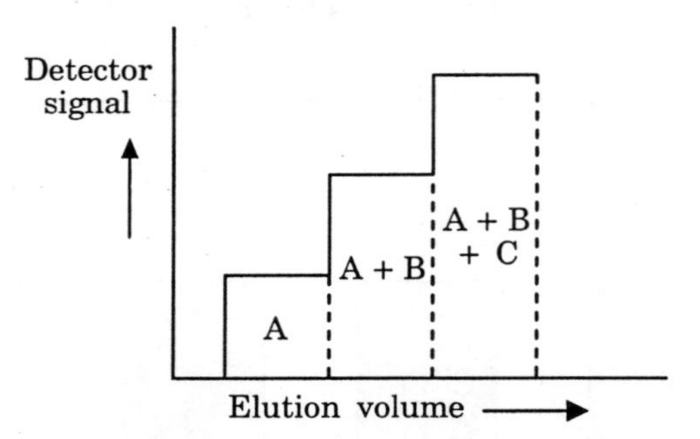

Figure 9.5 Frontal development.

Displacement analysis is practiced to regenerate the column after its use or to concentrate a particular component in the given mixture. A displacing agent D, with a stronger affinity for the stationary phase than the components A and B of the mixture, dissolved in the mobile phase is allowed to pass through the column after the introduction of the sample mixture. The displacing agent D displaces the less strongly held B from the column which in turn displaces the still less strongly held A and so on resulting in a elution profile as shown in Figure 9.6.

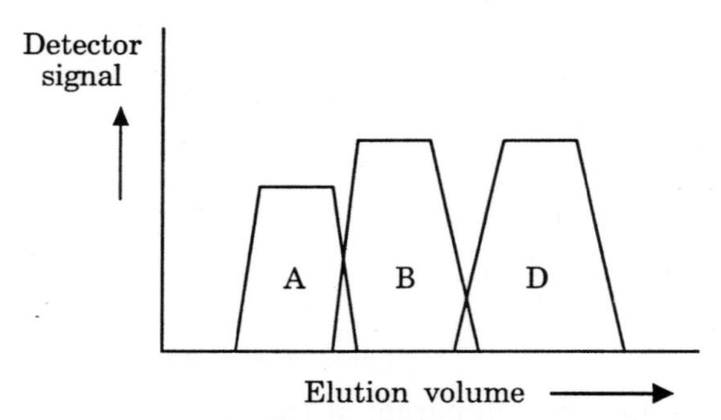

Figure 9.6 Displacement development.

9.4 CHROMATOGRAPHIC TERMS AND PARAMETERS

The various terms and parameters used in chromatographic techniques are exclusive and need explanation. It is also necessary to know the relationships between the various parameters to understand the theoretical principles, their significance and practice of chromatography.

9.4.1 Partition Coefficient

The plate theory of chromatography based on thermodynamic principles considers the chromatographic process as a series of transfers of the solute between the stationary and mobile phases. Based on the theoretical aspects developed for fractional distillation and countercurrent extraction, the concept of theoretical plates was introduced to describe the chromatographic process. The chromatographic system may be visualized to be composed of a number of discrete but contiguous layers or a series of plates called theoretical plates with the distribution of the solute between the two phases attaining equilibrium in each plate. The concept is inaccurate in that the chromatographic system never attains equilibrium as the solute is continuously exposed to either a fresh stationary phase or a fresh mobile phase. The concept is useful to predict the relative mobilities of different solutes on the basis of their partition coefficients and the Gaussian shape of chromatographic peaks but it fails to account for peak broadening.

The basis of all chromatographic techniques is the partition or distribution ratio or coefficient, which describes the way in which a compound or solute, which enters a chromatographic system, immediately distributes itself between the stationary and the mobile phases. The concentration of the solute in each phase is given by the partition coefficient, K, as given by Eq. (9.1).

$$K = \frac{C_s}{C_m} \tag{9.1}$$

The terms C_s and C_m are the molar concentrations of the solute in the stationary and the mobile phases respectively. Ideally, the partition coefficient is a constant over a wide range of concentrations, i.e., C_s is proportional to C_m. Chromatography in which this equation is applicable is called linear chromatography. If $K = 1$, the solute is equally distributed between the two phases. The value of K determines the average velocity of each solute zone, particularly at the zone center in the mobile phase. If two components in a mixture have values $K_A = K_B$, then separation factor $K_A/K_B = 1$ and no separation occurs. However, if $K_B > K_A$ then component A elutes out first and separation of the two components depends on the relative magnitudes of the partition coefficients.

9.4.2 Retention Time

A fundamental retention parameter in column chromatography is the retention time, t_R, The retention time is defined as the time taken by the solute to reach the detector from the moment of its injection into the column and may be determined by measuring the distance between the sample injection point to the apex of the peak from an on-line chromatogram as shown in Figure 9.4 and expressed in time scale. The retention time depends on the flow rate, F of the mobile phase, which is given by Eq. (9.2).

$$F = V \times \frac{P}{t_m} \tag{9.2}$$

The volume of the column, V is given as the product of the cross-sectional area of the cylindrical column, $\pi d^2/4$ multiplied by the length, L. The term P refers to the porosity of the stationary phase and t_m refers to the time required by a molecule of the mobile phase to pass through the column. The retention time of any solute that spends no time in the stationary phase ($K = 0$) i.e., a solute that is wholly unadsorbed, will be equal to t_m. The mobile phase flows through the column at a velocity, u, and the time it spends in the column, t_m is given by:

$$t_m = \frac{L}{u} \tag{9.3}$$

The corrected retention time $t_{R'}$ is given by Eq. (9.4).

$$t_{R'} = t_R - t_m \tag{9.4}$$

In gas chromatography where the mobile phase is a gas, temperature and pressure must be specified and retention volume must be corrected for the compressibility of the gas since the gas moves relatively slowly near inlet than at the exit end of the column.

The retention time of a solute may be related to the partition coefficient by expressing the migration rate of the solute, v, as a fraction of the velocity of the mobile phase, u and the fraction of time, t, the solute spends in the mobile phase as $v = ut$. This fraction is actually the average number of moles of the solute in the mobile phase at any given instant divided by the total number of moles of the solute in the column. The number of moles of the solute in the mobile phase is equal to $C_m V_m$ (i.e., concentration of the solute is the mobile phase volume of V_m). The total moles present is equal to the sum of the moles present in the mobile and stationary phases. Hence the above equation may be rewritten as:

$$v = u\frac{C_m V_m}{C_m V_m + C_s V_s} \quad \text{or} \quad v = u\frac{1}{1 + (C_s V_s / C_m V_m)} \tag{9.5}$$

Substituting K for C_s/C_m

$$v = u\frac{1}{1 + K(V_s/V_m)} \tag{9.6}$$

Since the migration rate of the solute is $v = L/t_R$ and $u = L/t_m$

$$\frac{L}{t_R} = \frac{L}{t_m} \frac{1}{1 + K(V_s/V_m)} \tag{9.7}$$

The retention time of the solute is related to the partition coefficient as given by

$$t_R = t_m \left[1 + K \left(\frac{V_s}{V_m} \right) \right] \tag{9.8}$$

The retention time of a solute is affected by various factors. These include: (i) nature of the stationary phase (ii) composition of the mobile phase (iii) column dimensions i.e. length and diameter and (iv) mobile phase flow rate. For a chosen combination of the stationary phase and mobile phase composition, the retention time of a solute increases with increasing length of the column and with decreasing flow rate of the mobile phase (i.e. solute spends more time in the column before eluting out). The retention time of a solute is independent of the quantity of the sample pulse introduced into the column and depends only on the nature of the sample. Hence it is useful in qualitative identification of the solute. The solute may be identified by comparing its retention time with that of the authentic solute (chromatographic standard) under identical conditions.

9.4.3 Retention Volume

Retention volume, V_R, may be defined as the volume of the mobile phase (in cm^3) required to transport (elute) a solute from the point of its injection into the column and its passage through the column to the detector. It is obtained by multiplying the retention time of a solute with the flow rate, F, of the mobile phase.

$$V_R = t_R \times F \tag{9.9}$$

The corrected or adjusted retention volume is given by:

$$V_{R'} = V_R - V_m \tag{9.10}$$

The retention time depends on the flow rate while retention volume does not depend on the flow rate. The volume V_m or V_0 is equal to the product $t_m \times F$ and is a measure of the total volume of the mobile phase contained within the column. It is often called the column dead space volume or hold-up volume or void volume or interstitial volume. The relationship between the retention volume and partition coefficient is given as:

$$V_R = V_m \left[1 + K \left(\frac{V_s}{V_m} \right) \right] \quad \text{or} \quad V_R = V_m + KV_s \tag{9.11}$$

9.4.4 Capacity Factor and Retention Ratio

The column capacity factor or retention factor is a measure of the retention of a solute component (not to be confused with loading capacity of the column). It is also called the solute partition ratio or mass distribution ratio, k' and is actually the normalized retention quantity used to describe the migration rates of solutes through the column. It is defined as the ratio of the total amount (distinct from concentration) of the solute present in the stationary phase to that in the mobile phase. It is related to the distribution coefficient as given by

$$k' = \frac{C_s V_s}{C_m V_m} \quad \text{or} \quad k' = \left(\frac{K}{\beta}\right) \tag{9.12}$$

where β is called the phase ratio and is equal to (V_m/V_s—the ratio of the volume of mobile phase to that of the stationary phase). The capacity factor is a measure of the time spent by a solute in the stationary phase relative to the time spent in the mobile phase. It is calculated from the retention time or retention volume of the solute as given by:

$$k' = \frac{t_R - t_m}{t_m} \quad \text{or} \quad k' = \frac{V_R - V_m}{V_m} \tag{9.13}$$

The retention time of the solute and capacity factor are also related to the length of the column and the linear flow velocity of the mobile phase as given by:

$$t_R = \frac{L}{u}(1 + k') \quad \text{or} \quad t_R = t_m(1 + k') \tag{9.14}$$

For practical purposes, the value of k' should be between 1 and 5 as the solute retention time increases with increasing value of k' with no retention if $k' = 0$.

Another normalized retention quantity is called the retardation factor or retention ratio, R_f, used in planar chromatography such as paper and thin layer chromatographic techniques. It is defined as the mobility of the solute relative to the mobility of the solvent or mobile phase and is related to the capacity factor k' as:

$$k' = \frac{1 - R_f}{R_f} \tag{9.15}$$

9.4.5 Relative Retention

The relative retention, α, (also called separation factor or selectivity factor) defines the ability of the chromatographic system to separate peaks for two solutes. It describes their differential migration rates and is determined by their distribution constants as given by

$$\alpha = \frac{K_2}{K_1} \quad \text{or} \quad \frac{k_2'}{k_1'} \quad \text{or} \quad \frac{t_{R2}'}{t_{R1}'} \quad \text{provided } t_{R2}' > t_{R1}' \tag{9.16}$$

Relative retention depends on the nature of the two phases and the column temperature. A proper choice of the stationary phase is necessary so that the relative retention of a pair of solutes is at least 1.05 or greater up to 2. Values of $\alpha > 2$ increase the time required for completing the chromatographic run.

9.4.6 Column Efficiency

The efficiency of the separation effected by the column is related to the width of the chromatographic peak. Ideally the chromatographic peak has a Gaussian shape. For an ideal Gaussian peak, the peak widths at the inflection point at half height and at the base of the peak are related in terms of the standard deviation as shown in Figure 9.7. A Gaussian peak is described by the equation

$$y = y_0\, e^{-x^2/2\sigma^2} \qquad (9.17)$$

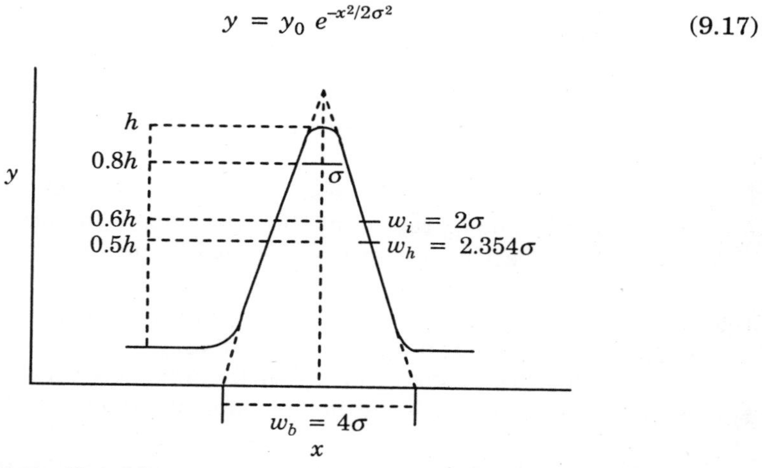

Figure 9.7 Gaussian nature of an ideal chromatographic peak.

where y is the height at any point on the curve and y_0 is the height at the peak maximum, x is the distance between the point on the curve and the ordinate passing through the peak maximum and σ is the standard deviation. The square of the standard deviation is variance. Since the Gaussian peak is symmetrical with respect to the ordinate, the peak width, w, at any point is $= 2x$. The inflection points ($x = \sigma$) of the peak occur at a height of 0.607 y_0 and the tangents to the peak drawn from the inflection points intercept the baseline at $x = 2\sigma$. Hence the peak width at the inflection points, w_i, at half height, w_h and at the base, w_b are given as:

$$w_i = 2\sigma; \quad w_h = 2\sigma\sqrt{(2 \ln 2)} = 2.354\sigma; \quad w_b = 2w_i = 4\sigma \qquad (9.18)$$

A convenient method for describing column efficiency on a quantitative basis involves the use of two related parameters, namely, the number of theoretical plates, n and the plate height, h. A Gaussian peak is described by its variance and the plate height is actually variance divided by the length of the column, $(h = \sigma^2/L)$.

The width of the peak depends on the number of theoretical plates n, increasing with $\sqrt{n}$. On the other hand the retention time increases directly with n. Thus σ is proportional to $\sqrt{n}$ while t_R is proportional to n. Therefore t_R/σ is proportional to $n/\sqrt{n}$ (i.e.= $\sqrt{n}$) and n is proportional to $(t_R/\sigma)^2$. Since $w_b = 4\sigma$ the value of n can be given as in Eq. (9.19) assuming the proportionality constant to be 1.

$$n = 16\left(\frac{t_R}{w_b}\right)^2 \tag{9.19}$$

The value n may also be calculated using width of the peak at half-height

$$n = 5.545\left(\frac{t_R}{w_h}\right)^2 \tag{9.20}$$

or

$$n = 6.28\left(\frac{t_R h_p}{a_p}\right)^2 \tag{9.21}$$

where w_h and w_b represent the peak widths at half height and the base of the peak and h_p and a_p represent the peak height and area of the peak respectively. It is easier to measure the peak width at half height. The widths, height and area must be measured in the same units as t_R. Usually they are measured in cm directly from the chromatographic chart paper using a ruler (and planimeter for area).

The length of the column occupied by one theoretical plate or one effective plate is defined as the height equivalent to a theoretical plate (HETP) or simply as plate height, h and is related to the length of the column (bed length), L and variance (σ^2) as given by Eq. (9.22).

$$h = \frac{\sigma^2}{L} \quad \text{or} \quad h = \frac{L}{n} \tag{9.22}$$

The efficiency of the column is also expressed in dimensionless quantity as the number of effective plates, N, using the adjusted retention time ($t_{R'}$) which is more representative of the actual retention of the solute as given by Eq. (9.23).

$$N = 16\left(\frac{t_{R'}}{w_b}\right)^2 \tag{9.23}$$

The terms N and n are related as given by Eq. (9.24).

$$N = n\left(\frac{k'}{1+k'}\right)^2 \tag{9.24}$$

As the value of k' increases, $(1 + k')$ is approximately equal to k' and the number of effective plates, N approaches the number of theoretical plates, n. The height equivalent to an effective plate (HEEP) or $H = L/N$.

The concept of theoretical plate is useful to characterize the performance of the chromatographic column. All symmetrical peaks in a chromatogram indicate approximately the same plate number. Both N and n are measures of the total efficiency of the column and depend on column length. The relative efficiencies of two columns of same or different lengths are compared in terms of number of plates/unit length by using the same compound for the experimental determination of retention time and peak width.

9.4.7 Resolution

Resolution or chromatographic resolution of two chromatographic peaks, R_s, is a measure of their separation and is defined as the distance between peak maxima compared with the average base widths of the two peaks. It is calculated by using the equation:

$$R_s = 2\left(\frac{t_{R2} - t_{R1}}{w_{b1} + w_{b2}}\right) \tag{9.25}$$

Retention time and peak width should be measured in the same unit to give a dimensionless value to resolution.

The chromatographic resolution depends on three independent factors namely, the column selectivity or separation factor (α), the retention factor or capacity factor (k') and the number of theoretical plates (n) and is given by the equation

$$R_s = \frac{1}{4}\left[\underbrace{\left(\frac{\alpha - 1}{\alpha}\right)}_{\text{Selectivity}}\underbrace{\left(\frac{k'}{1 + k'}\right)}_{\text{Capacity}}\underbrace{(N)^{1/2}}_{\text{Efficiency}}\right] \tag{9.26}$$

If $R_s = 1$ the two peaks just touch each other while for $R_s < 1$ the peaks are incompletely separated. A complete baseline separation of a pair of peaks requires $R_s \geq 1.5$. At this value the purity of the peak is 100%. It is necessary to mention here that a completely resolved peak is not equivalent to a pure substance. The peak may represent a series of solute components, which are not resolvable under the selected experimental conditions. The chromatographic run requires more time if $R_s \gg 1$ for all the pairs of peaks in a chromatogram. The selectivity factor is the most important. For example, a 1% change in the value of the selectivity factor represented by the first term in the equation (say from 1.01 to 1.02) will double the resolution. The effect of α on the first term is shown in Figure 9.8.

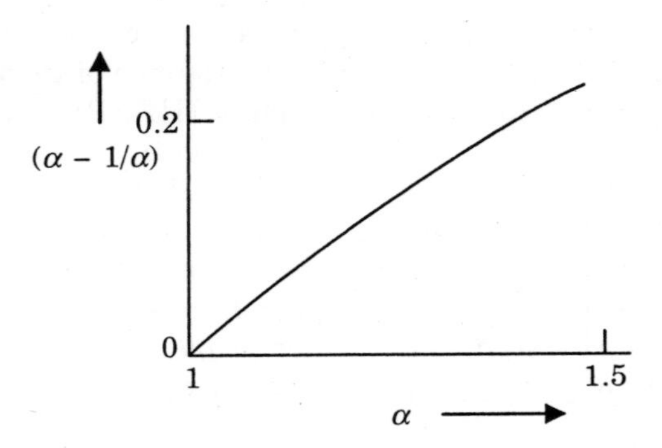

Figure 9.8 Effect of α on $(\alpha - 1/\alpha)$ term of the resolution equation. *Source:* J. Krijgsman, *Product Recovery in Bioprocess Technology*, BIOTOL series, Butterworth-Heinemann Ltd., Oxford, 1992.

The retention factor (second term in the equation) increases with k', as shown in Figure 9.9. If $k' > 5$ it will lead to slower separation as indicated by the plateau region in the Figure 9.9. However if $k' < 1$ resolution is low. The relationship between relative retention and the number of plates for achieving the same resolution is shown in Figure 9.10. As the number of theoretical plates increases with longer columns, resolution increases; but a four-fold increase in column length is necessary to double the resolution.

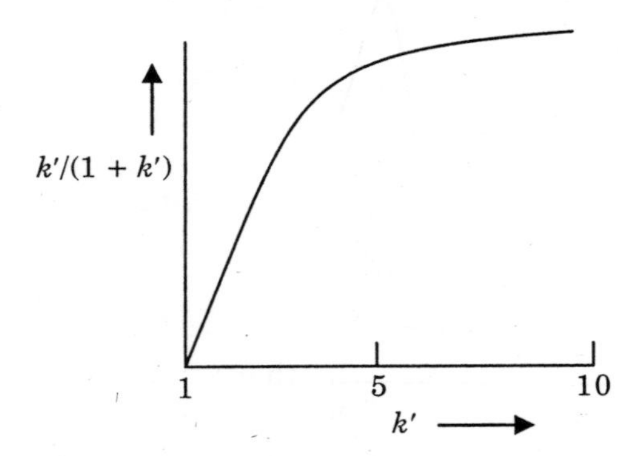

Figure 9.9 Effect of capacity factor on the $(k'/1 + k')$ term of the equation.

The effect of the three terms on the chromatographic peaks of a two component mixture is shown in Figure 9.11.

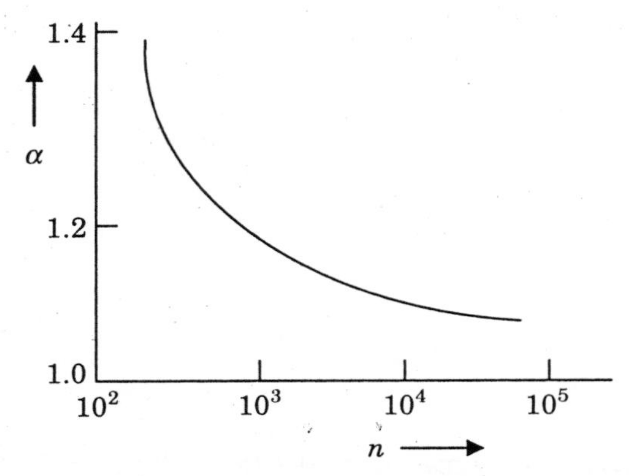

Figure 9.10 Relationship between selectivity and number of plates required for achieving the same resolution.
Source: J. Krijgsman, *Product Recovery in Bioprocess Technology*, BIOTOL series, Butterworth-Heinemann Ltd., Oxford, 1992.

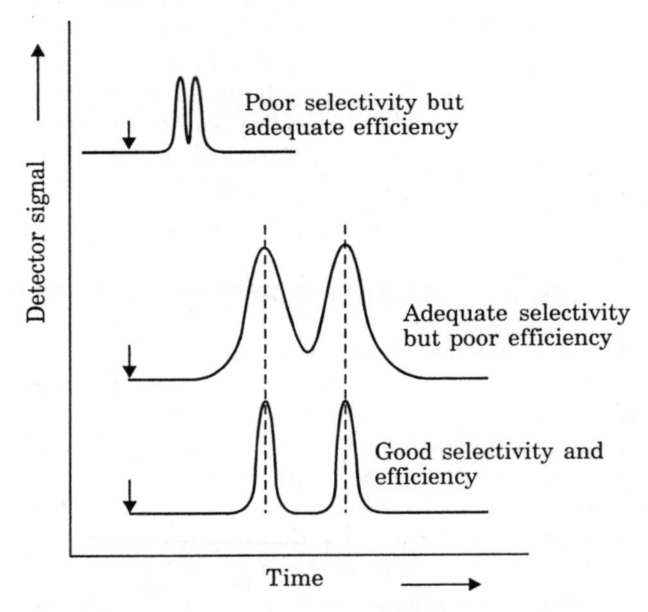

Figure 9.11 The combined effect of selectivity, capacity factor and efficiency on the separation of chromatographic peaks.

9.4.8 Peak Asymmetry

Asymmetrical chromatographic peaks are more common than perfectly Gaussian (symmetrical) peaks, the asymmetry being due to tailing or fronting (see Figures 9.12a and 9.12b). Asymmetry is more prevalent at lower concentrations mainly due to higher k' values or due to poorly

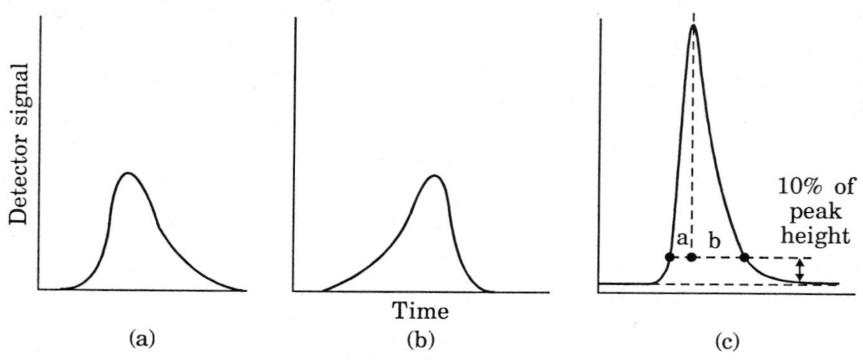

Figure 9.12 Asymmetric peaks (a) tailing (b) fronting and (c) determination of peak asymmetry.

packed columns and sample injection problems. Peak asymmetry can be a criterion of column performance and it is usual to quantify the degree of asymmetry or skewness. The asymmetry factor, A_s, may be defined as the ratio b/a, where b is the back half width and the term a is the corresponding front half width of the peak both measured at 10% peak height (see Figure 9.12c). The tailing factor, T_f, is given as $(a + b)/2a$. For a Gaussian peak, $a = b$ and both A_s and T_f are exactly 1. For most of the packed columns the asymmetry factor ranges between 0.90 (fronting) to 1.10 (tailing) and are considered acceptable.

9.4.9 Column Processes and Band Broadening

During elution of a solute through a chromatographic column, spreading and broadening of the elution band always occurs affecting the separation efficiency. The plate theory of chromatographic process is helpful in predicting the retention time or volume but is incapable of explaining the broadening of the band. The distribution of the solute between the phases is not instantaneous and requires a finite amount of time and hence a kinetic approach called the rate theory was developed to explain the broadening process. The rate theory as developed by van Deemter and others identifies the basic processes that contribute to the band broadening in a chromatographic column. These include the (i) tortuous and random path of the mobile phase (ii) longitudinal diffusion of the solute molecules (iii) resistance to mass transfer between the two phases.

These factors influence the width of the elution band and the effect of these factors of band broadening on the plate height may be explained as follows.

The plate height of a chromatographic column is a function of thermodynamic and kinetic processes that take place within the column. These include transverse and longitudinal diffusion of the solute, finite rate of equilibration of the solute between the stationary and mobile phases i.e., mass transfer, diffusion in the liquid stationary phase and flow irregularities leading to convective mixing. The relationship between plate height and the factors responsible for band broadening in liquid-liquid chromatography is expressed by van Deemter equation as:

$$h = A + \frac{B}{u} + (C_{\text{stationary}} + C_{\text{mobile}})u \tag{9.27}$$

The term A called the eddy diffusion, results from the inhomogeneity of flow velocities and path length of the mobile phase (and solute molecules) around the packed stationary phase particles. During the chromatographic run, some of the solute molecules pass through the column close to the column wall where packing density is low while other solute molecules pass through the more densely packed center of the column at a correspondingly lower velocity. As a consequence, molecules of the same solute elute out of the column at slightly different times leading to a broadening of the elution band. The magnitude of A depends on the particle diameter, d_p as given by the relationship $A = \lambda d_p$, where λ is an unspecified constant depending on the packing uniformity and column geometry. The magnitude of the term A may be minimized by decreasing the mean diameter of the particles, by a uniformly packed bed and a sufficiently higher pressure drop to force the mobile phase through the packed bed at a constant velocity. In the case of capillary columns there is no stationary phase particle packing and the term A disappears from the van Deemter equation.

The term B defines the effect of longitudinal or axial diffusion, i.e., the solute molecules diffuse from the concentrated center of the band to the more dilute regions both forward and backward from the band center within the mobile phase as it flows through the column. Longitudinal diffusion leads to peak dispersion or band spreading. The magnitude of B is equal to $2\gamma D_m$, where D_m is the solute diffusion coefficient in the mobile phase and γ is called the obstructive factor or tortuosity factor which indicates that longitudinal diffusion is minimized by uniformly dense packing and bed structure. The value of γ is equal to 1 for most of the coated capillary columns and 0.6 for packed bed columns. Solutes with high diffusion coefficient values cause elution bands to disperse axially along the column particularly at low mobile phase flow rates leading to band broadening. Band spreading is inversely proportional to mobile phase velocity because more time is available for diffusion at slower velocity of the mobile phase. The magnitude of B and hence, the band broadening, may be minimized by optimizing the flow rates of the mobile phase for a given column.

The term $C_{\text{stationary}}$ refers to the resistance to mass transfer at the solute-stationary phase interface. It is proportional to t^2/D_s, where t is the effective thickness of the stationary phase and D_s is the diffusion coefficient of the solute in the stationary phase. Slower rate of mass transfer due to low D_s values leads to peak broadening. The rate of mass transfer can be improved by decreasing the effective thickness of the stationary phase by choosing non-viscous liquids as the active stationary phase, which also give better D_s values.

The C_{mobile} term represents the radial mass transfer resistance. It is proportional to the square of the particle diameter of the packing and inversely proportional to D_m.

The peak broadening process may be minimized for a chosen chromatographic column by optimizing the linear flow rate of the mobile phase so as to determine the plate height minimum and achieve maximum separation efficiency. The effect of the various terms of the van Deemter equation individually and as a composite term on the plate height and the selection of optimum flow rates of the mobile phase are shown in Figure 9.13.

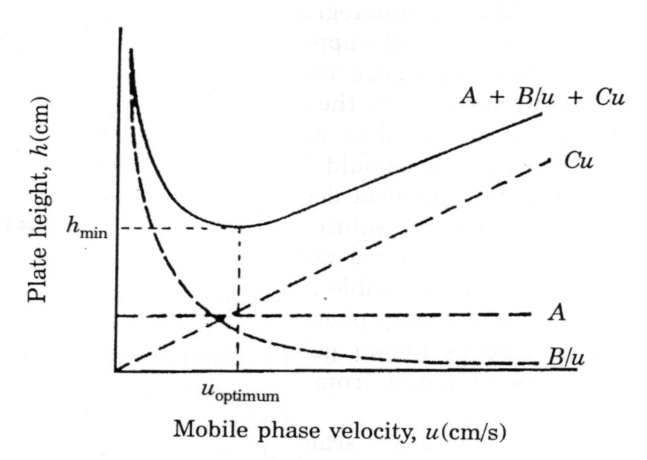

Figure 9.13 van Deemter plot indicating effect of velocity of the mobile phase on the efficiency of a given column.

9.4.10 Extra Column Broadening

The extent of band broadening is dependent to a major extent on the column. However, other factors also contribute to band broadening. These include the void volume or dead volume of the instrument, the injector design, sample volume and the flow properties of the detector, its volume and response time. The mobile phase flows through narrow connecting tubes between the injection port and the column and between the column and the detector. The tubing as well as the void space within the packed column, the injection port volume and the detector cell volume, all contribute to the void volume. The diameter and uniformity of the tubing is very critical since radial dispersion of the mobile phase along the cross-section of the tube causes band broadening. The extra column band broadening may be minimized by adopting appropriate measures. These include using short and narrow connecting tubes, connecting the tubes to the injection port, column ends and the detector through ferrules and nuts of very low dead volume, direct on-column injection, injecting a relatively small sample volume compared to the column volume and having a detector with a small cell volume.

9.5 PRACTICE OF CHROMATOGRAPHY

Stationary phases used in various chromatographic techniques vary widely. In fact, the choice of the chromatographic technique depends on the nature of the stationary phase. In the case of adsorption chromatography, inorganic solids such as silica, alumina, calcium carbonate are commonly used. In all other chromatographic techniques, the stationary phase is supported on inert solid supports such as silica, alumina or porous glass beads. Other examples include cellulose, polyacrylamide, agarose and dextran. In such cases, the support matrix material should be water-insoluble, physically rigid to withstand the operational stress, permeable and macroporous. It should be chemically stable but at the same time be amenable for chemical derivatization so as to modify the surface suitably for binding the solute components through different mechanisms. It should have minimal non-specific adsorption characteristics. The material should be reusable and preferably be of low cost.

The different types of stationary phases used in bioseparations include ion exchange resins, size exclusion gels, reverse phase, affinity and hydrophobic adsorbents prepared from synthetic polymer, agarose or cellulose matrices.

The *column* used in laboratory scale operations is mostly made of glass, acrylic plastic or polypropylene and of steel in the case of larger scale operations. The column material should be chemically inert, pressure- and solvent-resistant and preferably transparent. Most laboratory columns have 5–25 mm inner diameter and a length of 30–75 cm. The length of the column is chosen based on the resolution required. The stationary phase is packed into the column according to the manufacturer's instructions. Mostly wet packing of the stationary phase gels is practiced; wherein, a dilute slurry of the gel particles are pumped into the column at a constant flow rate. It is important to achieve uniform packing density throughout without introducing air pockets or void space. After packing to a desired length, the column is stabilized by running the solvent at a higher flow rate in order to obtain a homogenous and well packed zone at the head of the column. The stabilization of the column is necessary for perfect application of the sample. After packing the column the bed is washed with 4–5 column volumes of suitable solvent or equilibrating solvent prior to use.

The static capacity of the stationary phase gel depends on its packing density and its availability for interaction with a solute component and is measured in batch experiment, allowing sufficient time for equilibration. In contrast, the dynamic capacity of a gel is the binding capacity under operating conditions in a chromatographic column.

Mobile phases used in various chromatographic separations include pure solvents, binary and ternary mixtures of solvents, buffers and salt solutions. The chosen mobile phase should be compatible with the chosen stationary phase in that it should be chemically inert towards the stationary phase and also to the sample components. The sample should be readily soluble in the mobile phase.

Sample introduction is usually carried out by injecting a small amount of (pulse) sample with the help of a capillary or hypodermic syringe at the head of the column or a special arrangement called loop or adaptor. Solid as well as liquid samples are dissolved preferably in the mobile phase or some solvent compatible with the mobile and stationary phases.

Elution techniques include isocratic elution and step or continuous gradient elution (pH, ionic strength, concentration, polarity gradient). In HPLC, flow programming where the flow rate of the mobile phase is changed during chromatographic development, and in GC temperature programming, where the column oven temperature is programmed to increase at specified heating rates, are also possible.

Isocratic elution is the simplest elution technique used for the development of chromatography wherein all the experimental conditions, particularly the eluent composition is kept constant throughout the development. Under such a condition, the retention times of the components in a mixture and thereby the separation, depend on the partition coefficients of the individual components. Isocratic elution is suitable only to a narrow range of samples which can elute out at reasonable intervals of time giving rise to well resolved peaks. However, in most cases, resolution is not accomplished by isocratic elution and this problem is referred to as the general elution problem.

A *general elution problem* is often encountered in chromatography when mixtures containing widely disparate components need to be separated. If elution conditions are adjusted to obtain resolution acceptable to components with short retention times, then elution of components with greater k' values will be retained in the column for too long for a practical analysis. On the other hand, if elution conditions are adjusted to suit the resolution of components with long retention times, then components with shorter retention times may not get separated. A convenient method to overcome this problem is to change the elution conditions during the chromatographic development, the method being called gradient elution.

Gradient elution involves the change of eluent composition during the chromatographic development either continuously or stepwise so as to change the partition coefficient values of each component with time. The gradient is formed with respect to percentage composition of a mixture of eluents, or pH or ionic strength of the eluent. The elution technique is useful for samples with a wide range of capacity factor values, which are not easily separated under isocratic conditions. A mobile phase gradient is formed by mixing two or more solvents either incrementally (step-wise) or continuously. The most frequently used gradients are binary solvent systems with a linear, concave or convex increase in the percent volume fraction of the stronger solvent. The three gradient shapes can be expressed in simple equations as given below.

$$\text{Linear gradient} = Q_B = \frac{t}{t_g}$$

$$\text{Concave gradient} = Q_B = 1 - \left(1 - \frac{t}{t_g}\right)^n$$

$$\text{Convex gradient} = Q_B = \left(\frac{t}{t_g}\right)^n$$

where Q_B is the volume fraction of the stronger solvent, t is the time after the gradient started, t_g is the total gradient time and n is an integer for controlling the gradient.

9.6 HPLC

High performance liquid chromatography also known as high pressure liquid chromatography (HPLC) is a versatile and sophisticated method suitable for a wide variety of analytical and preparative applications. The instrument used for this consists of the main components already described in section 9.3, namely, (i) mobile phase delivery system (ii) high pressure pump (iii) sample injection port (iv) column (v) detector (vi) data processor and (vii) sample or fraction collector. The components are highly sophisticated as described below.

Mobile phase delivery system. The sub components of the mobile phase delivery system consist of solvent reservoir, solvent degassing unit, micro filter, microprocessor controlled high pressure pump, PTFE and SS tubing and mixing chamber. Two or three high pressure pumps are required when using gradient mixed solvent mobile phase. The solvent reservoir is a storage container of material resistance to chemical attack by the mobile phase (usually a closed glass bottle). A flexible PTFE hose fitted with a microfilter (2 μm pore size) connects the reservoir to the high pressure pump. The mobile phase should be free from any dust or particulate matter which can damage the pump or the column and hence the liquid is passed through a microfilter. The degassing unit is required to prevent gas bubble formation, which damages the column packing, degrades the pump and detector performance particularly when different solvents are mixed or mobile phase is depressurized. A further problem associated with dissolved oxygen in the mobile phase is the oxidative degradation of the sample and a reduction in the sensitivity and operating stability of UV, refractive index, fluorescence and electrochemical detectors. Degassing is carried out by applying vacuum above the solvent, by sparging with nitrogen or helium or better by ultrasonic treatment.

The microprocessor controlled *high pressure pump* is one of the most important components of HPLC since its performance directly affects the detector sensitivity and reproducibility of chromatographic separations. The criteria for a good HPLC pump include its capability to operate at pressures up to 500 atmospheres. It should be capable of precise and pulse-free delivery of the mobile phase at flow rates in the range 0.02 to 10 ml/min with minimum drift. The pump should have minimum response time (of the order of microseconds) for changing the flow rates. Two types of pumping systems are commonly used in HPLC, constant pressure pumps such as gas displacement and pneumatic amplifier pumps and constant flow pumps such as reciprocating and syringe pumps.

The gradient devices may be either low-pressure mixing or high-pressure mixing types and are required with a combination of a minimum of two pumps to perform gradient elution.

Sample injection device is a valve capable of introducing a wide range of sample volumes into the column as a sharp pulse or plug without adversely affecting the column. Since the system is operating at a high pressure, sample injection is achieved with a typical volume of 10–20 μl (sometimes as high as 100 μl) through a sample loop. While the sample is being injected by a hypodermic syringe, the mobile phase flows through the sample injection valve at high pressure directly to the column (eluent bypass). Then the valve is turned or rotated to allow the mobile phase to sweep the injected sample in the sample loop on to column at the same pressure.

Column is made of SS tubing 25 cm long and 3 to 4 mm diameter for analytical purposes. The precolumn is relatively small of about 8–10 cm long while HPLC columns used for preparative purposes are much wider. The columns contain the packed bed of stationary phase particles.

Support materials for column packing are small rigid particles mostly of silica or alumina (ceria, zirconia and thoria used less frequently) having a narrow particle size distribution. Three general packing materials are commonly used (i) superficially porous particles of about 30 mm size, (ii) very small totally porous particles of 5 mm size and (iii) totally porous particles of 10 mm size for the adsorption and partition mode of HPLC. Rigid, porous polymeric beads based on synthetic polystyrene crosslinked with divinyl benzene or crosslinked natural polymers such as dextran or agarose are used for size exclusion chromatographic mode of HPLC. Bonded phases prepared by reacting organo chlorosilanes or organo alkoxysilanes with porous silica particles are used in reverse phase mode. Ion exchange chromatography uses ion-exchange resins based on synthetic or natural polymeric gels. For affinity, pseudo-affinity, covalent and hydrophobic interaction chromatographic techniques, the stationary phase consists of specific ligands coupled or immobilized on agarose beads.

The stationary phase particles are packed to a very high density and are also of uniform density in order to achieve high efficiency for separation between components. The high-density packing results in minimum plate height and increase in number of plates per unit length of the column, which contribute to the enhanced separation efficiency. However, the resistance to the flow of the mobile phase through the bed of particles also increases necessitating the use of high pressure pump.

A short column called *pre-column* or *guard column* is placed between the sample injector and the main column to protect the main column from damage or loss of efficiency due to the presence of particulate matter or strongly adsorbed material in the sample. It also functions as a saturator column to prevent dissolution of the stationary phase of the main column. The guard column is usually packed with the same stationary phase as that of the main column but of coarse particles.

Two types of *detectors* are used in HPLC. A bulk or solvent property detector such as refractive index (RI) detector or solute property detectors such UV spectral detector, UV diode array detector, UV–Visible spectrophotometer, fluorescence detector or electrochemical detector may be used. The signal from the detector is amplified and fed to data processor system for visual display and printout of the chromatogram.

The *fraction collector* in most of the sophisticated HPLC systems are microprocessor controlled programmable units and are easily synchronized with detector signals to collect the components of the sample mixture separately.

Modes of operation: Chromatographic separation of components in a mixture depends on their having differing k' values which in turn depends on their having differing distribution ratios between the stationary and the mobile phases. The variety of stationary phases used in liquid chromatography results in a variety of separation modes. The versatility of HPLC is based on the ability of the instrument to be operated in different modes of chromatography simply by changing the column. These include adsorption chromatography, partition chromatography, reversed phase chromatography, ion exchange chromatography, ion pair partition chromatography, size exclusion chromatography and affinity chromatography. Except for the size exclusion chromatography, all other modes are essentially suitable for low molecular weight substances.

Adsorption chromatography is useful for the separation of non-polar or fairly polar organic molecules. During chromatographic run, the solute components are eluted out on the stationary phase (a bed of silica or alumina particles) with a non-polar organic mobile phase such as *n*-hexane or dichloromethane. The solute components are retained on the column according to their affinity for the adsorbent surface.

Partition chromatography uses a stationary phase consisting of an inert support (silica) coated with a layer of organic liquid (e.g., polyethylene glycol, β-,β-oxydipropionitrile) which is insoluble in the mobile phase. The mobile phase is non-polar *n*-hexane. Separation is based on the relative solubilities of the solute components in the two phases.

Reversed phase chromatography is the most widely used mode. Highly polar solutes give rise to problems of long retention and peak tailing in adsorption chromatography. The use of a non-polar stationary phase (reversed phase) in conjunction with a polar mobile phase overcomes such problems. The polarity of the stationary phase is reversed from highly polar to non-polar by covalent modification of the silica surface with hydrocarbon molecules. The most commonly used hydrocarbon bonded stationary phase is the octadecylsilane (ODS), in which the silica stationary phase particles are modified with 18 carbon atom hydrocarbon chain. The stationary phase known as ODS or C-18 or RP-18 is highly non-polar and widely used for a variety of organic molecules in conjunction with water/methanol mixture as the mobile phase. Similar relatively more polar C-8 (RP-8) and C-2 stationary phases are also used in conjunction with more non-polar/polar solvent mixtures as mobile phase. *Hydrophobic interaction chromatography* is similar to reversed phase chromatography which is specifically applicable to protein separation and purification.

Ion-exchange chromatography makes use of ion-exchange resins, both cationic and anionic resins, containing ions capable of being exchanged with ionic solutes in the mobile phase. *Ion pair partition chromatography* is a similar method for the separation of ionic or ionizable molecules.

Size exclusion chromatography involves the separation of solute components on the basis of their molecular size (molecular weights or shapes). The pore size of the stationary phase is such that smaller molecules have access to the pores while bigger molecules are excluded from the pores. The technique is useful in the separation of high molecular weight substances (polymers). Synthetic polymers require organic mobile phase and the technique is generally called *gel permeation chromatography*. Natural biopolymers are separated by size exclusion in aqueous media and the technique is termed *gel filtration*.

Affinity and pseudoaffinity chromatography uses biospecific or specific interactions between the chromatographic stationary phase and the biomolecules in the mobile phase for efficient separation/purification of biomolecules.

9.7 SCALE-UP OF CHROMATOGRAPHY

Scale-up in bioseparations varies with the nature of the product and may range from milligram to gram to kilogram scale. Chromatographic technique is highly amenable for changing the scale of operations from laboratory scale to pilot plant and large-scale level.

The restrictions imposed on using higher flow rates by the flow of the mobile phase around beads resulting in viscous forces as well as rates of diffusion in and out of beads of the stationary phase may be overcome by the use of membrane chromatography or radial flow column chromatography.

Membrane chromatography uses permeable membranes containing small pores with adsorbent material in the form of beads bound to the membrane surface that facilitate higher flow rates of the mobile phase even at relatively low operating pressures. A stack of many such membranes resembles a packed bed column but does not create diffusion problems as the adsorbent ligands are exposed readily to the solute components in the flowing mobile phase. Higher flow rates in the range of 100–500 cm/h are possible in membrane chromatography.

Radial flow column chromatography facilitates the flow of the mobile phase radially from the outside of the column cylinder to the inside to a collection tube located at the center the packed column as shown in Figure 9.14.

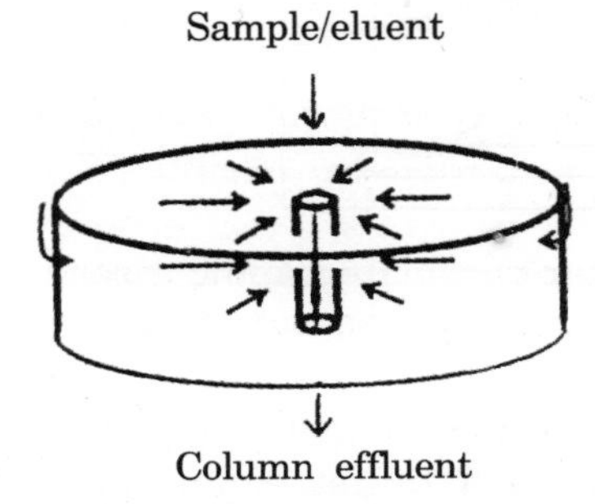

Figure 9.14 Radial flow chromatographic column.

Source: R.K.. Scopes, *Protein Purification—Principles and Practice*, 3rd ed., Narosa Publishing House, New Delhi, 1994.

The mobile phase flows in a perpendicular direction to the flow in normal column chromatography. The stationary phase consists of beads similar to that in normal columns but higher flow rates are attained because of the large cross-sectional area at the commencement of the contact with the stationary phase. Though the cross-sectional area decreases rapidly as the flow approaches the center, the pressure drop is negligible. The solute components get separated initially at a low linear flow rate but rush out quickly towards the end as they move from outside of the column to the center.

9.8 PLANAR CHROMATOGRAPHIC TECHNIQUES

Planar or two-dimensional chromatographic techniques include paper chromatography (PC) and thin layer chromatography (TLC). The techniques are mostly used for analytical purposes. However, TLC can be used for preparative work for isolating/purifying products up to a few hundred milligrams.

Paper chromatography uses a strip of filter paper in the place of column of column chromatography. The stationary phase is actually the immobilized water held by cellulose molecules of the filter paper and the mobile phase is a homogenous liquid. The paper chromatographic assembly is shown in Figure 9.15. The chromatographic chamber is a glass jar provided with a cover. The eluent reservoir such as a petri dish

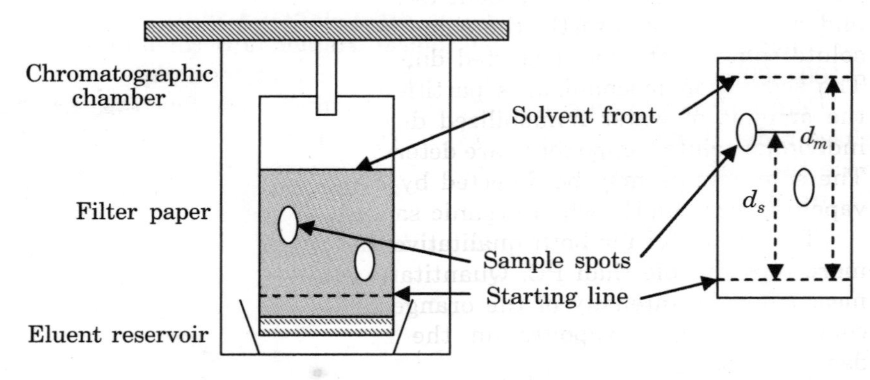

Figure 9.15 Planar chromatographic assembly and the paper strip after development.

or beaker containing the chosen mobile phase is placed inside the jar and the atmosphere is allowed to get equilibrated with the mobile phase vapour. A small amount of sample is spotted near one end of the filter paper strip with the help of a capillary or hypodermic syringe. The filter paper is placed inside the jar as shown in the figure to allow the chromatogram to develop. The mobile phase moves up the paper (ascending chromatography) by capillary action. The solute components of the sample also move up the paper depending on their partition coefficient values between the stationary and mobile phases. Different solute components of the mixture will move to different extents and get separated depending on their partition coefficient values. At a specified time when the solvent front has moved up wetting almost to 80–90% of the filter paper strip, the strip is removed from the chamber and the distances the solute components and the solvent have moved are noted. The movement of the solute is detected by spraying the paper with an appropriate chemical reagent to develop a colour or by exposing it to UV light to visualize the emitted fluorescence. The retardation factor, R_f, for a given solute is calculated as $R_f = d_s/d_m$, where d_s and d_m represent the linear distances of the solute and the mobile phase from the starting line as shown in Figure 9.15. The R_f values are analogous to partition coefficient or retention time values of solutes and range between 0 ($K = \infty$) and 1

(K = 0). Paper chromatography is mostly used for qualitative analysis and rarely for quantitative estimations.

Thin layer chromatography is similar to paper chromatography in experimental methodology except that the support for the stationary phase is a glass plate or aluminum foil. The plate is coated with fine particles of an adsorbent such as alumina or silica gel as a thin layer of about 0.2 mm thickness. The thin layer of adsorbent is formed by coating a slurry of the adsorbent in water and allowing the water to evaporate by drying at room temperature or at a higher temperature. Drying at room temperature allows the adsorbent to hold water. The experimental setup and procedure are exactly the same as for paper chromatography. The solute components get separated due to their different migration rates. The separation mechanism is partition and/or adsorption depending on the amount of water immobilized during drying. The R_f values of the individual solute components are determined as in paper chromatography. The solute spots may be detected by exposing the TLC plate to iodine vapours, particularly when organic samples are spotted.

TLC is useful for both qualitative and quantitative analysis as it is more reproducible than PC. Quantitative analysis can be carried out by measuring the intensity of the orange or brown colour produced by the condensing iodine vapours on the spots of organic solutes with a densitometer.

High performance thin layer chromatography (HPTLC) uses thin layer plates coated with very fine adsorbent particles of uniform size of about 5 μm diameter to a film thickness of about 100 μm. The high performance plates provide a sharp separation indicative of about 4000 theoretical plates per 3 cm requiring a development time of 10 min. compared to the normal plates exhibiting 2000 plates in 12 cm requiring 25 min. for development. However, the HPTLC plates have smaller sample capacity. HPTLC coupled with sophisticated densitometer or fluorescence detector is highly useful for quantitative analysis.

9.9 PROCESS CONSIDERATIONS IN PREPARATIVE LIQUID CHROMATOGRAPHY

Chromatographic processes find extensive use, both for analytical and preparative applications. There is little difference between analytical and preparative liquid chromatography as far as the overall chromatography is concerned, but the latter faces non-chromatographic problems such as product isolation or solvent recovery. The optimum performance of any chromatographic process has to consider four main performance characteristics, namely, capacity, recovery, resolution and speed, which may be considered as the corners of a tetrahedron. In general, optimization of any of these characteristics is achieved only at the expense of the others. For analytical estimations resolution and speed are the main criteria. The face of the tetrahedron joining speed, resolution and recovery is important and capacity is sacrificed in the case of analytical applications. Hence the use of small-scale equipment is the general rule.

In preparative applications, particularly in bioseparations, it is necessary to define the goal and fix the priorities such as whether the chromatographic process is for (i) isolation, concentration and stabilization of the desired product from a complex mixture (ii) purification of the desired product by eliminating bulk impurities or (iii) final polishing operation. Recovery is of great importance in any preparative operation particularly in the production of high value products. Recovery is affected by destructive interferences in the sample and also by unfavourable conditions in the column. It becomes more important further downstream as the value of the purified product increases.

In the isolation of the desired product from a crude feed, capacity and speed are more important than resolution in order to handle large volumes, keep the scale of equipment small and reduce the cycle time. In such a situation, selective binding capacity for the desired product in the presence of contaminants is the critical parameter. Selective binding will also allow stepwise elution of the product thereby concentrating the product. High speed is another critical parameter in order to minimize sample application time so that the desired product is not destabilized or denatured by contaminants (for instance, proteins may undergo denaturation or proteolysis).

If the preparative chromatographic process aims at purification of the isolated product, a trade off between resolution and capacity for high productivity (recovery) becomes necessary. Resolution becomes more important followed by the capacity to produce the product per unit volume of chromatographic media in unit time. Resolution is enhanced by selective binding of the product and more importantly by selective desorption of the product from the column. Selective desorption is achieved by continuous gradient or a multi-step elution procedure. High speed operation is not critical as the desired product has been stabilized as well as concentrated in the isolation step.

Finally, if the preparative chromatographic process aims at polishing the product, resolution is the most important criterion because polishing aims at achieving the required quality/purity of the product by removing trace contaminants such as structural variants of the product, host proteins, endotoxins, nucleic acids, viruses and reagents. In such a situation, high resolution is achieved by using selective as well as high efficiency media with small bead size and selective desorption from the column by changing the shape and slope of the continuous gradient elution technique.

Exercises

1. Calculate the retention volumes, capacity factors and relative retention of two solutes giving peaks 1 and 2 with retention times of 3.5 and 5.0 minutes respectively on an adsorption column at a flow rate of 2.2 ml/min. The t_0 value is 1.5 min.

Solution

$$V_{R1} = t_{R1} \times F = 3.5 \times 2.2 = \mathbf{7.7} \text{ ml}$$

$$V_{R2} = 5.0 \times 2.2 = \mathbf{11.0} \text{ ml}$$

$$V_m = 1.5 \times 2.2 = \mathbf{3.3} \text{ ml}$$

$$k'_1 = \frac{t_{R1} - t_m}{t_m} = \frac{3.5 - 1.5}{1.5} = \mathbf{1.33}$$

$$= \frac{V_{R1} - V_m}{V_m} = \frac{7.7 - 3.3}{3.3} = \mathbf{1.33}$$

$$k'_2 = \mathbf{2.33}$$

$$\alpha = \frac{t'_{R2}}{t'_{R1}} = \mathbf{1.75}$$

$$\frac{k'_2}{k'_1} = \mathbf{1.75}$$

2. A solute X of a sample mixture is not retained on the column and elutes out at 4 minutes after injection. Another solute in the same sample mixture Y elutes out at 12 minutes. The mobile phase flow rate is 20 ml/min. Calculate the values of V_m, k' for the solute Y and the time spent by Y in the mobile and stationary phases.

Solution

$$t_m = 4 \text{ min and } V_m = t_m \times F = 4 \times 20 = 80 \text{ ml}$$

From Eq. (9.13), $k' = (t_R - t_m)/t_m$
k' for the solute Y = 2
Since t_m = 4 min, the time spent by solute Y in the stationary phase is **8** minutes and the time spent by Y in the mobile phase is **4** minutes.

3. A chromatographic separation of a two component samples on a 50 cm column gave the retention times for the solutes A and B as 2.5 and 3.1 minutes with base widths of the two chromatographic peaks being 0.24 and 0.3 minutes respectively. Calculate the (a) number of theoretical plates, (b) plate height and (c) resolution of the two peaks.

Solution

(a) Using Eq. (9.19), $n = 16(t_R/w_b)^2$ for the two peaks individually, the number of theoretical plates would be:

$$n \text{ for peak A} = 16 \times \left(\frac{2.5}{0.24}\right)^2 = \mathbf{1736}$$

$$n \text{ for peak B} = 16 \times \left(\frac{3.1}{0.3}\right)^2 = \mathbf{1708}$$

$$\text{Average value of } n = 1722$$

(b) Using Eq. (9.22) $h = L/n$

Plate height, h = 50 cm/1722 plates = **0.029 cm/plate**

(c) Using Eq. (9.25), $R_s = 2[(t_{R2} - t_{R1})/(w_{b1} + w_{b2})]$

$$R_s = 2\left(\frac{3.1 - 2.5}{0.24 + 0.3}\right) = \mathbf{1.11}$$

This value of 1.11 is not sufficient to achieve base line resolution.

4. Using the data in exercise 3 calculate the resolution of the two peaks on an 80 cm column.

Solution

Calculate the new retention times, number of theoretical plates and base widths of the two peaks on a 80 cm column.

From Eq. (9.14), t_R is directly proportional to L provided the flow rate and capacity factor do not change with the longer column.

$$\text{Hence } t_{R1} = 2.5 \times \frac{80}{50} = 4.0 \text{ min and } t_{R2} = 3.1 \times \frac{80}{50} = 4.96 \text{ min.}$$

Since the number of theoretical plates is also directly proportional to L, values of n can be calculated for the two peaks on the longer column as:

$$n = 1736 \times \frac{80}{50} = 2778 \text{ and } 1708 \times \frac{80}{50} = 2733$$

The corresponding base widths of the two peaks are 0.30 and 0.38 min respectively

The resolution of the two peaks on the 80 cm long column = **2.8**.

The resolution is adequate for baseline separation as $R_s > 1.5$.

5. In order to determine the optimum flow rate of the mobile phase for a 50 cm column containing uniform packing of 30 μm particles of stationary phase trial experiments were conducted using a solute at different flow rates of the eluent. The data obtained for three different linear velocities of the eluent are as follows:

S.No.	u (cm/min.)	t_R (min.)	w_b (min.)
1	10	7.5	0.4
2	30	2.5	0.13
3	50	1.5	0.09

Determine the optimum flow rate of the mobile phase for the column and calculate the number of theoretical plates for optimum conditions of flow.

Solution

Calculate the number of theoretical plates and plate height for the three flow rates. The values are 5625, 5917 and 4444 plates/cm and 8.88×10^{-3}, 8.45×10^{-3} and 1.12×10^{-2} cm/plate respectively for the three linear flow rates of the mobile phase of 10, 30 and 50 cm/min.

Using van Deemter Eq. (9.27) $h = A + B/u + Cu$, we have

$$0.0088 = A + \frac{B}{10} + 10C$$

$$0.00845 = A + \frac{B}{30} + 30C$$

$$0.0112 = A + \frac{B}{50} + 50C$$

Now we calculate the coefficients A, B and C. The values are $A = 1.096 \times 10^{-3}$ cm, $B = 0.0598$ cm^2/min and $C = 1.788 \times 10^{-4}$ min respectively.

Since the term A involves the particle size of the stationary phase particles which remain constant for a given column, A can be taken as constant. Then,

$$h \sim \frac{B}{u} + Cu$$

The most efficient column should have the smallest value for h_{min}. So setting $dh/du = 0$

$$h = \frac{B}{u^2} + C$$

and $u_{opt} = (B/C)^{1/2}$ and $h_{min} = A + 2\ (B/C)^{1/2}$

$u_{opt} = 18.3$ cm/min and h_{min} at optimum flow rate of **0.007636 cm/plate**.

The number of theoretical plates at optimum conditions of flow rate = **6548**.

Questions

1. Explain the terms (i) partition coefficient (ii) retention time (iii) retention volume (iv) capacity factor (v) relative retention (vi) resolution (vii) plate height and (viii) number of theoretical plates. What is their significance?
2. Explain the terms involved in van Deemter equation and their significance.
3. Describe the column chromatographic process.
4. Give a schematic diagram of the chromatographic set-up and describe the functions of the components.
5. Write a note on the chromatographic development techniques.
6. Give an account of the different operating modes of HPLC.
7. Discuss the principle involved in and the practice of TLC.

CHAPTER

10

Gel Filtration

10.1 SIZE EXCLUSION CHROMATOGRAPHY

The use of column chromatography for separation of proteins on the basis of difference in their size by passing them down a column containing swollen particles of a gel of maize starch was first described in 1955. The flow rate for elution was very low because of the swollen particles. Porath and Flodin used cross-linked dextran as the matrix which enabled them to use relatively higher flow rates. They coined the term *gel filtration* for the separation technique based on molecular size. A number of such gels are available at present and their composition and pore size are carefully controlled so that molecules over a relatively narrow size range can now be separated by choosing an appropriate gel. The gel filtration technique was later extended to non-aqueous media by the introduction of polystyrene based matrices and the technique was called *gel permeation chromatography* (GPC). At present, the universally accepted name for separations based on molecular size is size exclusion chromatography and includes both GPC and gel filtration.

10.2 BASIC PRINCIPLES

The principle of gel filtration may be illustrated with reference to Sephadex, one of a number of commercially available materials. The gel consists of polysaccharide dextran cross-linked to form a three-dimensional network in the shape of beads. The cross-linked dextran is highly polar because of its large content of hydroxyl groups and swells considerably when placed in water. The extent of cross-linking during manufacture of the gel controls the pore size within the gel beads. Small molecules can enter the pores of the gel but larger molecules are excluded as shown in Figure 10.1. Thus the accessible volume of solvent is very much less for molecules totally excluded from the gel than for small molecules which are free to penetrate the gel as shown in the figure.

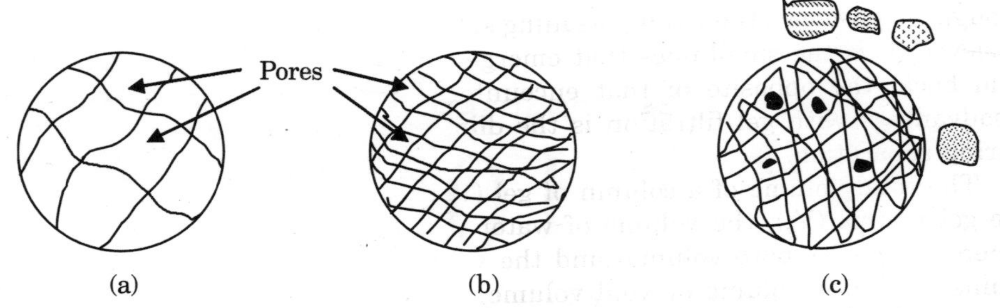

Figure 10.1 Pore size variation in beads: (a) large pores (b) small pores and (c) exclusion of large sized molecules (●) and inclusion of small molecules (●) from pores.

The separation may be regarded as due to the different amount of time the different solutes stay within the liquid phase that is entrapped in the matrix. This time is related to the fraction of the pores that are accessible to the solute. Experimentally the mixture of small and large solute molecules is placed on top of the column and eluted with a suitable eluent. As the solute molecules pass down the column, the small molecules diffuse into the gel and follow a longer path than the large molecules, which are completely excluded, from the gel particles. Eventually complete separation occurs, with the large molecules leaving the column first and the smallest one last. The schematics of separation by gel filtration are shown in Figure 10.2. The method is sometimes called *molecular sieving,*

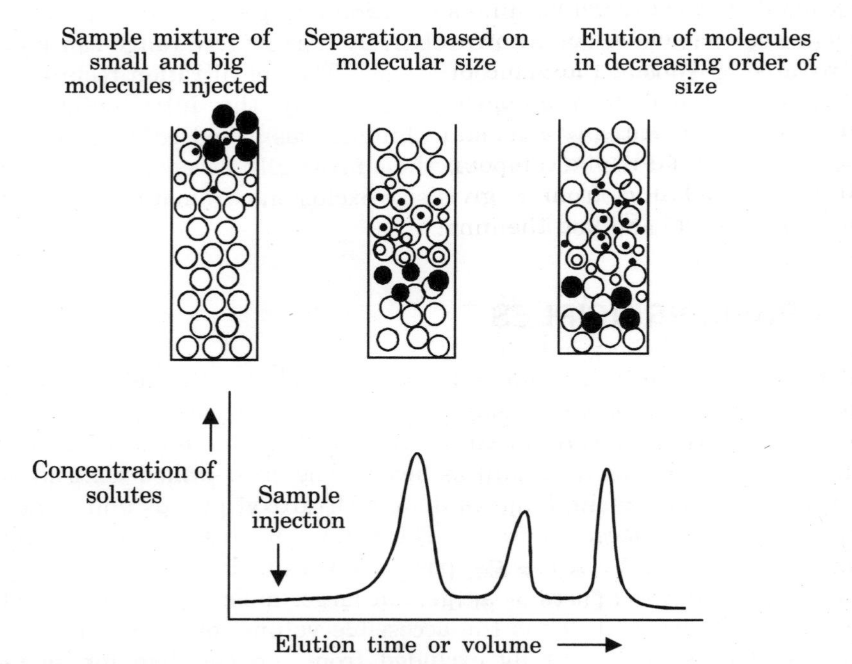

Figure 10.2 Separation of solutes by size exclusion in gel filtration column and the corresponding chromatogram.

though this conveys the wrong meaning since, when a mixture of particles is sieved, it is the small ones that emerge first while the larger ones are held back, the opposite of that encountered in gel filtration. A slight disadvantage with gel filtration is the dilution of the solute that occurs during separation.

The total volume of a column of gel (V_t) is the sum of the volume of the gel matrix (V_g), the volume of water inside the gel particles V_i (the inner volume, or pore volume) and the volume of water outside the gel grains V_0 (outer volume or void volume).

$$V_t = V_0 + V_i + V_g \tag{10.1}$$

The void volume is the volume of the liquid required to elute molecules that are completely excluded from the gel. This is determined by chromatographing a high molecular weight substance which cannot enter the gel pores. The solvent volume which leaves the column between the start and appearance of this substance in the effluent corresponds exactly to the volume of liquid between the gel particles. The total volume of the gel bed is determined by measuring the volume of the column from the support of the gel bed to the upper end. The inner volume of the gel is calculated from a knowledge of the dry weight of the gel (W) and the water regain (S) or solvent regain.

$$V_i = WS \tag{10.2}$$

Water regain or solvent regain is defined as the grams of solvent taken up by one gram of xerogel (a gel with no solvent or dispersing agent). If the dry weight of the gel is not known then the inner volume may be calculated by substituting the value of V_g by the density of the swollen state, d, in Eq. (10.1). The density of the gel in the wet state is determined by pycnometer and represents a constant for each gel in a given solvent.

The elution volume V_e of a compound is the volume of eluent required to elute that compound from a given column. It is related to the distribution coefficient, K_d and the inner and outer volumes of the gel as given in Eq. (10.3).

$$V_e = V_0 + K_d V_i \tag{10.3}$$

The distribution coefficient indicates the fraction of the inner volume accessible to a particular compound and is independent of the geometry of the column. The distribution coefficient is related to the selectivity of the column material. The selectivity of a gel filtration medium is *not* adjustable by changing the composition of the mobile phase unlike in other types of chromatographic media such as ion-exchange or reverse phase matrices.

Since $K_d = (V_e - V_0)/V_i$ as per Eq. (10.3), three possibilities occur in gel filtration of a mixture of solutes of different molecular size or weight:

1. For large molecules which are completely excluded from the gel, $K_d = 0$ and $V_e = V_0$.
2. For molecules of very small size which have complete accessibility to the gel pores, $K_d = 1$ and $V_e = V_0 + V_i$.
3. For molecules of intermediate size K_d will vary between 0 and 1. If however, $K_d > 1$ then adsorption of the compound on the gel has occurred and such a situation has to be avoided.

For achieving complete separation of the components in a given mixture, the sample volume to be applied on a given column of specific volume is limited by the difference in the elution volumes of the solutes. Thus for a two component mixture, the sample volume, V_s that can be applied to a column is given by Eq. (10.4).

$$V_s = V_{e1} - V_{e2} \tag{10.4}$$

However, in practice, the sample volume applied should be less than V_s because elution curves spread beyond ideal limits due to diffusion and flow irregularities in the column.

A plot of K_d versus log M (molecular weight of the proteins) or a plot of elution volume versus log M, yields a sigmoid selectivity curve as shown in Figure 10.3.

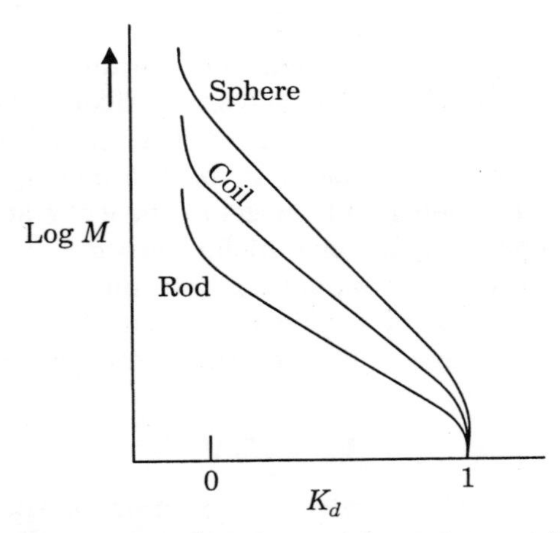

Figure 10.3 The selectivity curves for solutes with different shapes.

The middle range of the selectivity curve may be approximated to a linear relationship given by Eq. (10.5),

$$K_d = a - b \log M \tag{10.5}$$

The value of b (the slope of the selectivity curve) depends on the width of the pore size distribution of the column material and the value of the intercept, a is a function of the mean pore size. The sigmoid shape of curve reduces the practical working of the gel to about $0.1 < K_d < 0.9$.

The selectivity may also be given in terms of relative retention as in Eq. (10.6).

$$\alpha = \frac{V_{e2}}{V_{e1}} \tag{10.6}$$

Resolution R, in gel permeation chromatography may be expressed as in Eq. (10.7).

$$R = \frac{\alpha - 1}{4\alpha} \frac{k'}{1 + k'} \sqrt{N} \tag{10.7}$$

Resolution will be doubled if the number of theoretical plates N is increased by four times keeping α and k' (capacity factor) constant. Alternatively resolution may be improved by increasing α or k'. However, maximum resolution can be achieved by maintaining the condition $1 < k' < 5$.

10.3 RELATIONSHIP BETWEEN MOLECULAR SIZE OF SOLUTE AND RETENTION IN GEL FILTRATION

The retention of solutes in an ideal gel filtration, is governed solely by the differences between solute dimensions and pore dimensions. Hence gel filtration provides a means of determining the molecular weight or size (Stokes radius) of native or denatured globular proteins under a wide variety of conditions of pH, ionic strength and temperature.

The relationship between molecular size and weight of macromolecular solutes is strongly dependent on solute shape and may be expressed in terms of proportionality between the radius of gyration R_g of the solute and its molecular weight M as in Eq. (10.8) where a is a constant depending on the shape of the molecule.

$$R_g \propto M^a \tag{10.8}$$

The value of a is 1 for rod shaped molecules, 0.5 for flexible coils and 0.33 for spherical molecules. The shape of the molecule may be obtained by comparison of the hydrodynamic measurements (intrinsic viscosity, sedimentation or diffusion coefficients) with the radius of gyration from light scattering experiments. The radius of gyration is proportional to the viscosity radius of spherical solutes and flexible polymers but not to that of rigid macromolecules. The use of a well-defined polymer such as dextran to calibrate the column according to hydrodynamic volume is a better way to characterize an unknown solute by gel filtration.

Calibration of a Sephadex column by plotting K_d versus protein Stokes radius for the accurate (within ±5% accuracy) estimation of molecular weight of proteins in detergent solutions is possible. The Stokes radius, R_{St}, is defined as the radius of a sphere that would have the same frictional coefficient as the protein. The molecular weight of a compact globular protein is given by

$$\log M \sim 0.147 \pm 0.041 + \log \left(\frac{R_{St}^3}{v} \right) \tag{10.9}$$

where v is partial specific volume, if unknown, may be assigned a value of 0.74.

The size of solutes may also be obtained from viscosity data yielding an expression for the hydrodynamic volume V_h as given in Eq. (10.10).

$$V_h = \frac{\eta M}{\gamma N} \tag{10.10}$$

Where η is intrinsic viscosity, N is Avogadro number, γ is known as Simhas factor which depends on the shape of the molecule (2.5 for spheres and >2.5 for ellipsoids). For flexible polymers and spherical solutes, the hydrodynamic viscosity radius R_h is calculated using Eq. (10.11).

$$R_h = \left(V_h \frac{3}{4} \pi \right)^{1/3} \tag{10.11}$$

The relationship between viscosity radius R_h (in Å) and K_d is similar to the sigmoid shaped curves obtained for the selectivity curves shown in Figure 10.3.

The hydrodynamic radius is also related to the molecular weight of globular proteins and globular solvated proteins as given by Eqs. (10.12a) and (10.12b) respectively.

$$R_h = 0.718 M^{1/3} \tag{10.12a}$$

$$R_h = 0.794 M^{1/3} \tag{10.12b}$$

The conformation of proteins may be normalized to random coils to avoid ambiguities involved in less well-defined shapes. Denaturing media such as 6–8 M guanidine hydrochloride promotes disruption of the non-covalent bonds and the native protein reverts to the state of random polymer coils. Disulphide bonds of proteins may also be broken by reductive cleavage with mercaptoethanol or dithiothreitol and subsequent oxidation prevented by carboxymethylation. Low concentrations of detergents such as SDS may be used for denaturation of proteins. The hydrodynamic property of the protein-SDS complex indicates it to be rod-like.

10.4 MATERIALS FOR GEL FILTRATION

Cross-linked dextrans (Sephadex) and agarose (Sepharose), polyacrylamide and porous glass gels are commercially available. The degree of cross-linking is carefully controlled to give a range of products capable of fractionating molecules over a limited size range. The useful separation range of molecules is only approximate since separation depends on the size as well as shape and to a minor extent on the charge on the molecules. Sephadex is stable between pH 4 and 10 and temperature of 0° to 30°C. It is unaffected by dilute acids or bases but concentrated solutions of acids hydrolyse the glycosidic linkages. A preservative such as toluene, phenol or chloroform is added when storing the gel in a wet condition to avoid bacterial growth. Sephadex can be repeatedly used and after use washed with water and stored in a cold room for quite a time. The fractionation range of Sephadex and Sepharose gels are shown in Table 10.1.

TABLE 10.1 Fractionation Range of Commercial Gels

Gel		*Range of globular proteins (daltons)*	*Range of dextrans (daltons)*
Sephadex	G 10	Upto 700	Upto 700
	G 15	1,500	1,500
	G 25	1,000 – 5,000	100 – 5,000
	G 50	1,500 – 30,000	500 – 10,000
	G 100	4,000 – 150,000	1,000 – 100,000
	G 200	5,000 – 600,000	1,000 – 200,000
Sepharose	4B	60,000 – 20,000,000	30,000 – 5,000,000
	6B	10,000 – 4,000,000	10,000 – 1,000,000

10.5 EQUIPMENT

Gel filtration is mostly conducted in chromatographic columns. The equipment consists of a column packed with the gel beads. A pulse of the sample (about 3–5% of the gel volume) preferably dissolved in the eluent is introduced at the head of the column. The constant supply of the eluent is brought about either by a pump or simply by a solvent reservoir, which is placed at an appropriate height to facilitate gravity flow.

Hose connections from the solvent reservoir to the column and from the exit end of the column to the detector and fraction collector is usually flexible silicone tubing. The detector is a solute property detector, mostly a UV detector with a selected wavelength filter or even a spectrophotometer. The components and the setup are the same as used for column chromatography and described in Chapter 9.

10.6 THE PROCEDURE

The gel filtration medium is chosen based on the solute characteristics and the requirement. The gel is packed in a column to have a bed of uniform packing density to avoid band broadening. The length of the packed column will affect both the resolution and time required for eluting out the sample. In general, resolution is proportional to the square root of the column length while elution time is proportional to the length of the column. The volume of the packed column is a measure of the column capacity and the sample volume should not exceed 5% of the column volume to achieve the desired separation. The sample should not be viscous. Since the separation depends only on the solute molecular size and not on the eluent composition, it is usual to use buffers which do not affect the structure and the biological activity of the solute. The ionic strength of the eluent is kept at 0.15 or greater to prevent undesirable ionic interactions between the solute molecules and the gel matrix. Lower flow rates of the eluent are preferable for achieving better resolution. After use, the column is repeatedly washed with the eluent buffer and finally with a buffer in which the gel can be stored.

10.7 APPLICATIONS

10.7.1 Fractionation

The ability of gel filtration technique to separate macromolecules on the basis of small differences in size is well established. The most favourable situation is when it is possible to use a gel that will exclude the protein of interest and include contaminants or vice versa. This requires the use of a gel of optimal properties of high pore volume, narrow pore size distribution and suitable pore size to elute the protein of interest at $K_d \sim 0.2$–0.4. The purification of antiIgE-β-galactosidase conjugate on a Superose-6 gel (operating molecular weight cut-off range 5–10,000 kDa) is an example. With proper experimental design it is possible to separate a desired protein from others differing in molecular weight by a factor of two or less as illustrated by the separation of IGF-1 (molecular weight = 7600) from its fusion partner ZZ (molecular weight = 14500) and uncleaved fusion protein on Superdex 75 HR 10/30 column using 0.25 M ammonium acetate buffer (pH 6).

10.7.2 Separation of Monomers from Dimers and Higher Aggregates

High resolution gel filtration provides a gentle means of separating proteins with a high degree of homogeneity containing dimers and higher aggregates. The technique is thus useful as a polishing step in a multi-step complex purification scheme particularly in the case of recombinant proteins and other proteins in industrial processes where protein aggregates must be reduced to below a specified level in the final product. The final purification of a mouse monoclonal antibody, IgG2a, from the mixture containing IgG dimer and transferrin on Superdex 200 has been reported. Figure 10.4 shows the separation of the oligomers (M_n), trimer (M_3), dimer (M_2) from the monomer (M_1) of bovine serum albumin (BSA) on a Sephadex G-150 column.

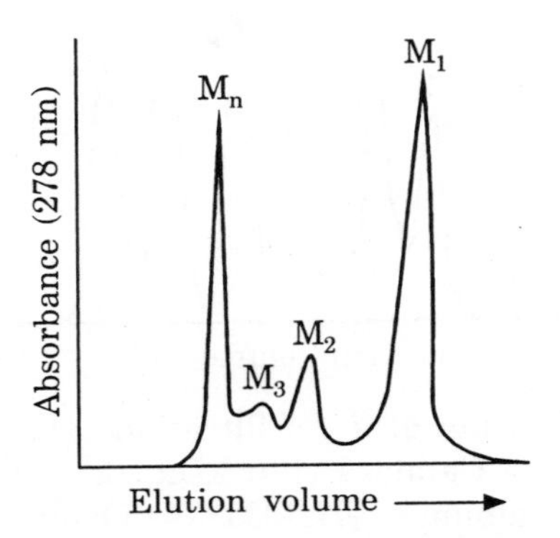

Figure 10.4 Separation of monomer from oligomers by gel filtration.

10.7.3 Estimation of Molecular Weight

A rough estimate of the molecular weight of proteins may be obtained from gel filtration experiment. Unlike electrophoretic techniques, gel filtration provides a means of determining the molecular weight or size (Stokes radius) of native or denatured globular proteins under a wide variety of conditions of pH, ionic strength and temperature. Gel filtration in the presence of urea or guanidine hydrochloride which transform polypeptides and proteins to a random coil configuration reducing structural differences is highly useful for molecular weight determinations. In practice, a mixture of molecular weight markers or standards (substances with known molecular weights) within the molecular weight cut-off range of the chosen gel is applied to a given gel filtration column and the chromatograms obtained. The test sample containing a protein whose molecular weight is to be determined is then eluted under identical conditions of eluent composition, flow rate and column dimensions. Based on the elution volume of the test sample, a rough estimate of the molecular weight can be obtained. An elution profile using protein standards for calibrating Sephadex G-200 superfine with potassium phosphate buffer (0.05 M, pH 6.8) containing 0.1 M NaCl and 2% sodium azide is shown in Figure 10.5.

Figure 10.6 shows the relation between molecular weight of proteins and their elution volumes on a Sephadex G-200 column (2.5 × 50 cm) at pH 7.5.

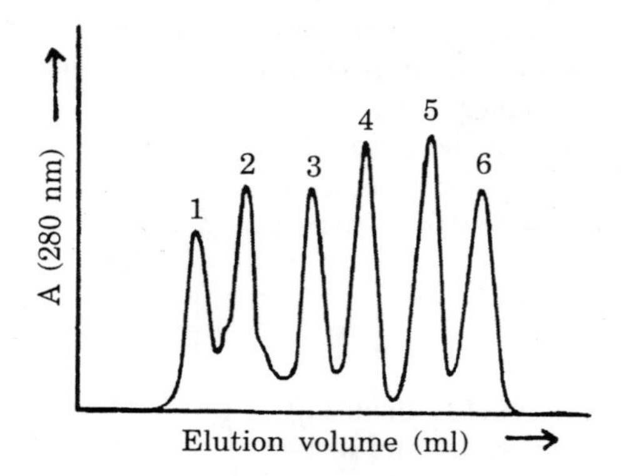

Figure 10.5 Calibration of a gel filtration column by chromatographic standards. (1) Catalase (170 kDa) (2) Aldolase (130 kDa) (3) Bovine serum albumin (67 kDa) (4) Ovalbumin (44 kDa) (5) Chymotrypsinogen A (21 kDa) (6) Ribonuclease A (14 kDa).
Source: Gel Filtration—Theory and Practice, Pharmacia Biotech.

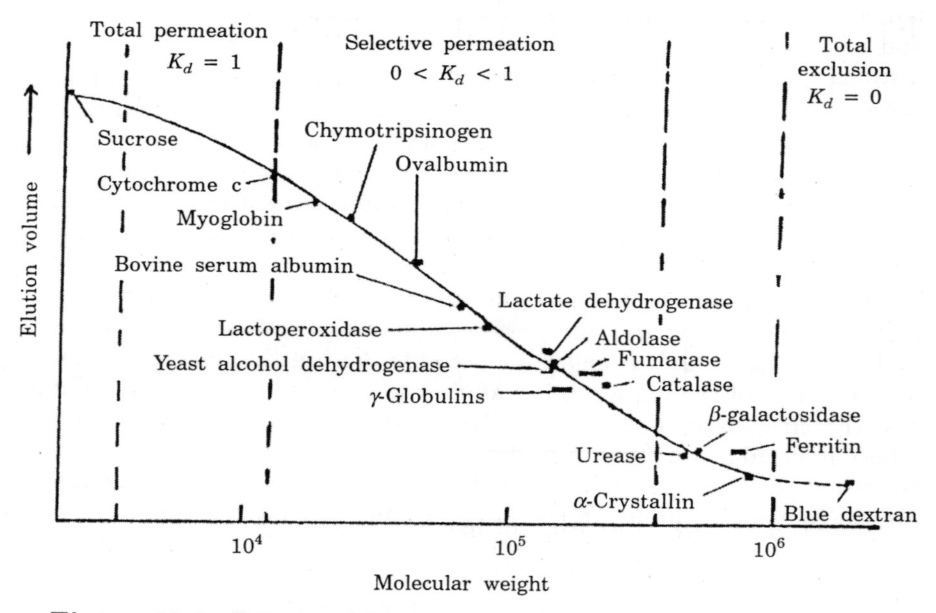

Figure 10.6 Relationship between elution volume and molecular weight for proteins on a Sephadex G-200 column at pH 7.5.

10.7.4 Determination of Molecular Weight Distribution of Polymers

Molecular weight distribution is an important characteristic of natural and synthetic polymers. The distribution analysis is easily done by gel filtration. The elution curve is recorded continuously or determined by investigating individual fractions of the polymer. It is not necessary to determine the molecular weight of the polymer in the effluent fractions separately as the chromatographic behaviour of the gel filtration column is highly reproducible. In addition, a calibration curve once determined for the column can be applied to a large number of runs.

10.7.5 Desalting

Gel filtration is highly useful for desalting as the biomacromolecules differ greately in size from salts and other small molecules particularly in laboratory experiments involving buffer exchange, enzyme assays and reactions, reactions that require adjustment of ionic strength and as a preliminary step prior to the use of ion-exchange chromatography, hydrophobic interaction chromatography and affinity chromatography.

The porosity of the gel is selected to exclude the solute to be desalted and the particle size is chosen to give low surface area. Low surface area is necessary to minimize adsorption and back pressure. The buffer salts and additives should be volatile or desalting should be carried out with distilled water. The profile for desalting haemoglobin on Sephadex G 25 is shown in Figure 10.7.

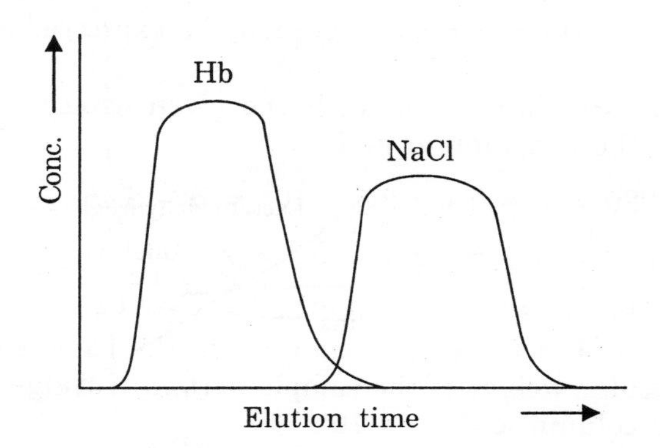

Figure 10.7 Desalting of haemoglobin by gel filtration.
Source: Gel Filtration—Theory and Practice, Pharmacia Biotech.

The technique is also useful for the removal of phenol from preparations of nucleic acids, removal of unincorporated nucleotides during DNA sequencing, removal of free low molecular weight labels (e.g., removal of ^{125}I, from solutions of labelled proteins), termination of reactions between macromolecules with low molecular weight substances, removal of products, cofactors and inhibitors from enzymes and removal of phenol red from culture fluids prior to anion exchange chromatography.

10.7.6 Determination of Equilibrium Constants

Gel filtration can be used to study chemical equilibria. In the case of slow reactions, where reactants and products can be separated on a gel filtration column, these substances can be quantitatively determined in the effluent for determining the equilibrium constant. In the case of fast reactions, one of the reactants is chromatographed in an eluent containing the other reactant. From the elution curve of the reactants, the reaction can be studied. This method is useful in the study of protein binding of low molecular weight substances such as drugs, to study the competition of two molecules for the same site and for estimation of reaction rates.

Exercises

1. Two proteins of molecular weights 2.5×10^5 and 1×10^4 were eluted out of a gel filtration column at 220 ml and 300 ml respectively. Determine the molecular weight of a protein which elutes out at 270 ml from the column under the same conditions.

   Solution

   The linear portion of Figure 10.6 may be expressed as $V_e = a - b \log M$.
   Two equations are set up with the given experimental data to calculate the constants a and b.

   $$220 = a - b \log 2.5 \times 10^5 = a - 5.4b \quad (1)$$

   $$300 = a - b \log 1 \times 10^4 = a - 4b \quad (2)$$

   The calculated values: $a = 528.55$; $b = 57.14$
   Substituting the values of a and b into the same equation gives the molecular weight of the sample protein eluting out at 270 ml from the column as 3.35×10^4.

2. A gel filtration column was calibrated using protein standards. The molecular weights of the standards and the corresponding elution volumes are 130 kDa (145 ml), 67 kDa (175 ml), 44 kDa (200 ml) and 21 kDa (225 ml). A sample protein eluted out at 210 ml from the column under identical conditions. Calculate its molecular weight.

   Solution

   From the graphical plot of log M versus V_e the value of log M is read off for V_e = 210 ml as 4.48 and the molecular weight of the sample protein is 30.2 kDa.

Questions

1. Explain the basic principle of gel filtration.
2. How are the inner and void volumes of a gel filtration column determined?
3. What is selectivity curve? What is its significance?
4. How is molecular weight of a protein determined by gel filtration?
5. Give an account on the applications of gel filtration.

CHAPTER

11

Ion Exchange Chromatography and Chromatofocusing

11.1 PRINCIPLE

Ion exchange chromatography (IEC) is a widely used technique for protein purification because of its versatility, high resolving power, high capacity and simplicity in operation. Protein separation and purification on the basis of charge is used in ion exchange chromatography as well as in chromatofocusing, electrophoresis and isoelectric focusing.

The basic principle in IEC involves reversible competitive binding of different ions of one kind to immobilized ion exchange groups of opposite charge bound to the chromatographic matrix called the ion exchanger. Thus for example, when a sample mixture containing components of positive charge is applied to a column containing cationic (negatively charged) matrix, solute components with more positive charges (M^{2+}) are adsorbed (exchanged) strongly compared to those with less positive charges (M^{+}). Components with no net charge (M^{0}) or a net negative charge (M^{-}) pass through the column unretained. The bound species may be eluted out giving rise to a chromatogram as shown in Figure 11.1.

The interaction between the proteins and the ion exchanger depends on several factors such as the net charge and charge anisotropy that is, charge distribution on the protein surface, ionic strength and pH of the solvent, nature of ions and other additives, for example, organic solvents present in the solvent. The higher the charge on the protein, the more strongly it will bind to a given ion exchanger of opposite charge. Similarly a given protein binds strongly to a highly charged ion exchanger, that is, one with a higher degree of substitution of charged groups. The pH of the medium or solvent is the most important parameter, which determines protein binding, as it determines the effective charge on both the protein

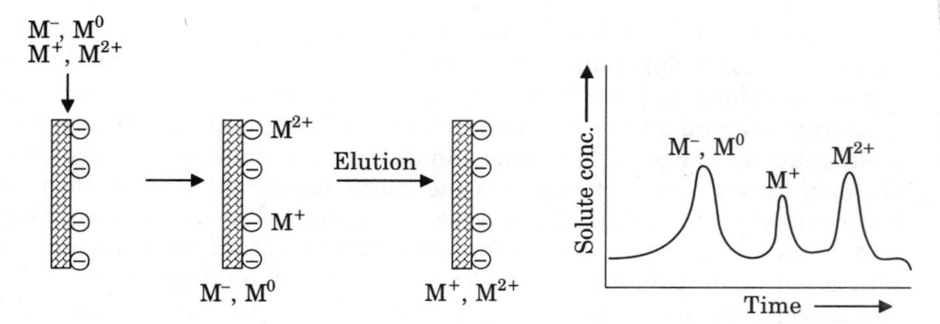

Figure 11.1 Schematic representation of the principle of IEC.

and the ion exchanger. Hence control of pH is essential to obtain reproducible results. In general, protein binding to an ion exchanger occurs only when the protein has a net charge of opposite sign to that on the ion exchanger. Thus at pH values far away from the pI of the proteins, they will bind strongly and will not desorb from the column at low ionic strength of the solvent. Protein binding is weak when the pH is closer to its pI. The distribution of charge on the surface of the protein also has an important role to play. Since the charges on the protein surface are distributed unevenly, weak binding can still occur near the protein pI even when the net overall charge is the same as that of the ion exchanger as shown by phosphoglycerate kinase and horse heart cytochrome oxidase. These two enzymes bind to a cation exchanger even at pH values where their net charges are slightly negative as they have clusters of residual positive charges exposed on the surface.

11.2 ION EXCHANGERS

The ion exchanger consists of a water insoluble matrix to which charged groups have been covalently bound. The charged groups are associated with mobile counter ions which can be reversibly exchanged with other ions of the same charge without altering the matrix. The charged groups may be either acidic or basic functional groups. Cation exchangers have acidic groups with a net negative charge on the matrix and positively charged exchangeable counter ions. Anion exchangers have basic functional groups with a net positive charge on the matrix and negatively charged exchangeable counter ions. The charge on the ion exchanger depends on the pH of the solvent. Four types of matrices are commonly available for use in IEC. These are (i) synthetic hydrophobic polymeric resins of hydrophobic polystyrene or partly hydrophobic polymethylmethacrylate crosslinked with divinylbenzene, (ii) naturally occurring as well as synthetic hydrophilic polymers such as cellulose, dextran or agarose, (iii) synthetic hydrophilic polymers made into hard beads for use in HPLC and (iv) silica gel.

The cationic functional groups attached to the matrix include carboxylic acid, sulphonate, sulphoethyl, sulphopropyl, carboxymethyl or ortho phosphate groups. Similarly anionic functional groups such as quaternary amine and diethylaminoethyl (DEAE) groups are attached to yield anion exchangers. Sulphonic and quaternary amino groups are used to form strong ion exchangers while other functional groups such as carboxymethyl and DEAE form weak ion exchangers. The terms strong and weak are relative and refer only to the extent of variation of ionization with pH and not to the strength of binding. Strong ion exchangers are completely ionized over a wide pH range. Weak ion exchangers exhibit a varying degree of ionization and also exchange capacity with change in pH.

All cationic exchangers have a limiting pH below which they cannot be used. In general the dissociation constant, pK, of the functional group is suggested as the lower limit of pH for use and is given in Table 11.1 for common functional groups. Similarly all anion exchangers have an upper limit of pH as indicated by the pK of the functional group beyond which they cannot be used. However, quaternary amines do not deprotonate even at high pH values and hence do not have an upper limit for use.

TABLE 11.1 Ion Exchangers

Name	*Symbol*	*pK*	*Functional group*
	Cation exchangers		
Carboxymethyl	CM	3.5–4.0	$-OCH_2COOH$
Orthophosphate	P	3 and 6	$-OPO_3H_2$
Sulphonate	S	2	$-OCH_2SO_3H$
Sulphoethyl	SE	2	$-OCH_2CH_2SO_3H$
Sulphopropyl	SP	2.0–2.5	$-OCH_2CH_2CH_2SO_3H$
	Anion exchangers		
Diethylaminoethyl	DEAE	9.0–9.5	$-OCH_2CH_2NH(C_2H_5)_2$
Trimethylhydroxy-propyl	QA		$-OCH_2CH_2CH(OH)N(CH_3)_3$
Quaternary aminomethyl	Q		$-OCH_2N(CH_3)_3$
Triethyl aminomethyl	TEAM	9.5	$-OCH_2N(C_2H_5)_3$
Polyethyleneimine	PEI		Polymerised $CH_3CH=NH$

11.3 CAPACITY OF ION EXCHANGERS

The capacity of an ion exchanger is a quantitative measure of its ability to take exchangeable counter ions. The *total ionic capacity* depends on the number of charged functional groups per gram of dry ion exchanger or per ml of swollen gel. It is usually determined by titration with a strong acid or base. Ion exchangers have a high degree of ionic capacity in the range of 100–500 μM/ml of bed. The *available capacity* of the ion exchanger is the amount of the charged solute that can be bound to gel under specified experimental conditions. The available capacity is referred to as the *dynamic capacity* in IEC under specified flow rate of the mobile phase through an ion exchange column. The available capacity of an ion exchanger is determined by batch method. In this method a series of solutions of different concentration of a given protein are added to a known quantity of ion exchanger in test tubes, equilibrated at suitable pH and ionic strength for binding to occur. After adsorption, the supernatants are assayed for protein concentration and the available capacity of the ion exchanger is calculated. The dynamic capacity of the ion exchanger in a column is determined by saturating the given volume of the ion exchanger in a column (usually 1 ml) with the protein solution at a specific flow rate, washing the column to remove excess of protein and then eluting the bound protein with appropriate buffer. The total amount of the bound protein eluted out gives the dynamic capacity of the ion exchanger at the specified flow rate.

The total capacity of an ion exchanger for binding proteins (protein binding capacity) is usually expressed in terms of serum albumin for anion exchangers and lysozyme or haemoglobin for cation exchangers. For example, CM-cellulose with a degree of substitution of total ionic capacity of 190 μM/ml binds 130 mg (2 μM) of BSA per ml of ion exchanger while for Q the corresponding figures are 300 μM and 65 mg of BSA (1 μM) per ml. Thus only a fraction of the total ionic capacity is used for protein binding under actual chromatographic conditions because of kinetic aspects. The dynamic capacity of ion exchanger is usually in the range of about 20% of the total ionic capacity.

The available and dynamic capacities depend critically on the properties of the ion exchanger matrix, the properties of the solute protein and experimental conditions. The properties of the ion exchange matrix include the porosity and the type and number of charged functional groups. High available capacity is obtained with a macroporous matrix highly substituted with ionic groups, which maintain their charge over a wide range of pH. Greater the pore size of the ion-exchanger, greater will be its dynamic capacity for a given protein. Thus, DEAE Sephadex A-50 with larger pores binds 250 mg of haemoglobin per ml of bed while DEAE Sephadex A-25 with smaller pores binds only 70 mg/ml of haemoglobin. Non-porous matrices have lower capacity than porous matrices. However non-porous matrices have higher efficiency due to shorter diffusion distances.

The properties of the solute protein which are of importance in determining the dynamic capacity of the ion exchanger are its molecular size/weight and its charge/pH relationship. The capacity of an ion exchanger is thus different for different proteins. More of small molecular weight proteins is bound to a given ion exchanger compared to large molecular weight proteins. Thus, CM-Cellulose binds 130 mg/ml (9 μM) of chicken egg white lysozyme (14,300 Da) and only 21 mg/ml (0.15 μM) of lactate dehydrogenase (140 kDa).

The experimental conditions which influence the dynamic capacity of the ion exchanger include the pH and the ionic strength of the buffer, the nature of the counter ion, the flow rate of the mobile phase and the temperature. The dynamic capacity depends critically on the flow rate of the mobile phase, decreasing with increasing flow rate.

11.4 OPERATING MODES

The amount of ion exchanger to be used should be about ten times that of the protein binding capacity. The ion exchanger may be used for batch adsorption or for column chromatography or expanded bed adsorption. In the batch separation process, a given quantity of ion exchanger previously equilibrated with an appropriate buffer is stirred with a solution of the sample till equilibrium is attained. The slurry is filtered and the ion exchanger is washed with buffer. In case of incomplete adsorption, the filtrate is stirred with a fresh batch of ion exchanger to strip the filtrate of all the desired solute. The filtered ion exchanger is then stirred with elution buffer of about 1–2 times the volume of the gel until the desorption process is complete. The slurry is filtered by suction to recover the desorbed product. Alternatively, the elution of the desired product from the gel may be carried out by packing the gel into a column. Batch process, though less efficient compared to column chromatography, is a rapid technique. The method is useful for separating the desired product from gross contaminants in the initial stages of bioseparation scheme as well as for concentrating it. The method is ideal for handling large volumes of sample containing low concentrations of the desired products and also when the sample viscosity is high which will lead to back-pressure and consequent clogging or fouling of the column.

Expanded bed adsorption technique is based on the principle of fluidization of the adsorbent bed (see Figure 11.2). The adsorbent bed

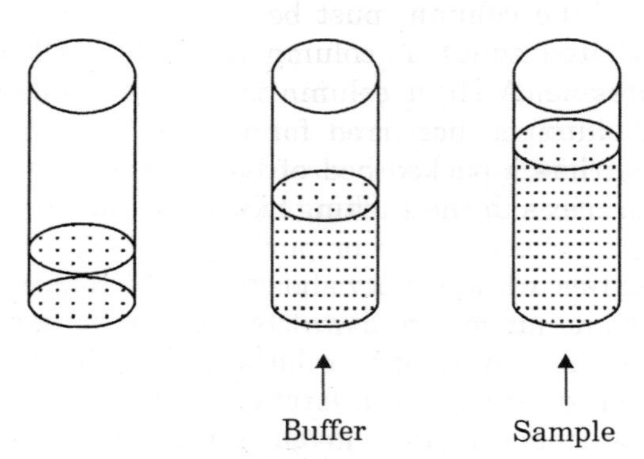

Figure 11.2 Expanded bed ion exchange adsorption.

expands as the adsorbent particles are raised by an upward flow of the liquid. The difference between a fluidized bed and expanded bed is that in the expanded bed there is little or no back-mixing of the particles. The adsorbent particles are kept in suspension by balancing the sedimentation velocity of the particles by the upward flow of the liquid. As the bed expands with the upward flow of the liquid, the movement of particles is very small. A homogeneous expanded bed and a constant velocity liquid flow are achieved by a unique design of the column and the choice of suitable adsorbents. The column has a special flow distributor at the bottom and the adsorbent particles have a well-defined size and density distribution.

11.5 PRACTICE OF ION EXCHANGE CHROMATOGRAPHY (IEC)

IEC has the capacity and the ability to adsorb and concentrate solute components from dilute solutions. In bioseparations it is a highly useful technique in the early stages of a separation scheme to concentrate and purify proteins. A proper choice of the stationary phase, mobile phase and experimental parameters such as pH, ionic strength, stepwise or gradient elution and presence or absence of additives is important in carrying out IEC.

11.5.1 Column Material and Stationary Phase

In column chromatography, the material of construction of the column should be chemically inert so as to prevent destruction of biological substances and also minimize non-specific binding of solutes. The bed support should be such that it facilitates free flow of the liquid with minimum clogging and should be easily replaced or exchangeable when necessary. The column should be capable of withstanding the back-pressure developed particularly when high performance medium is used. The dead space of the column must be kept to a minimum to prevent remixing of separated zones. A column length of about five times the diameter is preferable. A short column is preferred for gradient elution while a longer column is preferred for isocratic elution. In general, a laboratory column has a packed bed of height of 5–15 cm.

The ion exchanger and the starting buffer should be chosen depending on the requirement of separation or purification. The amount of the ion exchanger should be such that only about 10–20% of its dynamic capacity is utilized during the process so as to achieve maximum resolution. The chosen ion exchanger is washed thoroughly with water and equilibrated to the pH and ionic strength of the starting buffer containing the counter ion to be used during elution. The ion exchanger is packed into the column and the packing is tested for any irregularities of lumps or air bubbles. The number of theoretical plates and plate height may be determined by using a test substance to arrive at the optimum operating conditions.

11.5.2 Mobile Phase

The pH and ionic strength are the two important parameters influencing the property of the mobile phase. The charge on the protein is very sensitive to changes in pH and hence buffer capacity to maintain the required pH is important. The buffer salt concentration used is in the range of 0.01–0.05 M. Volatile buffers are preferred because direct lyophilization of the pooled fraction after chromatography can be carried out. For a cation exchanger, the buffering ion should be negatively charged, for example, for phosphate, carbonate, acetate or morpholine-sulphonate when the counter ion is ammonium, sodium or potassium. Ammonium acetate has been widely used for cation exchangers as it has the additional advantage of volatility. For an anion exchanger, the buffering ion should be positively charged, for example, tris buffer with chloride as the counter ion.

A non buffering salt such as sodium chloride is usually added to the mobile phase to maintain ionic strength so that proteins can be eluted out quickly. Additives may also be used particularly to enhance protein solubility by diminishing hydrophobic interaction, for example, ethylene glycol, ethanol, urea and sometimes detergents. Detergents must be neutral or carry the same charge as the ion exchanger.

11.5.3 Sample Preparation and Application

The amount of sample should be less than about 20% of the dynamic capacity of the bed and the sample volume preferably less than 5% of the bed volume. The sample is dissolved preferably in the starting buffer so that the ionic compositions of both are the same. The sample should be clear and free from any suspended particles or turbidity. The viscosities of the eluent and the sample should be similar as high viscosity of the sample causes zone mixing and irregular flow pattern. The sample may suitably diluted and in the case of nucleic acid samples, viscosity may be reduced by digestion with endonuclease. The recommended method of sample application is to inject it with a syringe onto the column or through a loop. The sample may also be introduced onto a drained bed just after draining the equilibrating buffer or may be introduced through a capillary at the top of the column containing a small amount of the eluent. After the application of the sample, the column is washed with starting buffer till no free or unbound components are available in the column.

11.5.4 Elution

A variety of elution techniques may be adopted. These include (i) isocratic elution (ii) stepwise elution (iii) gradient elution by change of pH or ionic strength or both (iv) affinity elution and (v) displacement chromatography. The components collected by elution are lyophilized and the column is washed thoroughly and regenerated as per the procedure given by the manufacturer of the ion exchanger.

11.5.5 Column Regeneration

The column is regenerated after each cycle by salt solution until the ionic strength reaches about 2 M to wash out all bound substances. The salt solution should contain the counter ion to the ion exchanger to facilitate equilibration. In some applications denatured proteins or lipids may remain in the column even after regeneration of the column necessitating cleaning-in-place (CIP) procedures. The CIP procedure involves the removal of tightly bound impurities by washing the column with 0.5–1.0 M sodium hydroxide solution, which removes the contaminants by dissolution. Sodium hydroxide is also useful in sanitization of the column for inactivating microbes.

11.5.6 Applications

IEC finds extensive use in the separation, concentration and purification of a variety of biomolecules, particularly labile substances. In the initial stages of a bioseparation scheme where sample dilution and contamination by impurities are quite common IEC is an ideal technique. In IEC solutes bind to the matrix gel at low ionic strength and elute out of the column at higher ionic strength whereas in hydrophobic interaction chromatography (HIC) binding is strong at high ionic strength and elution occurs at low ionic strength. Hence it would be advantageous to use these two techniques in a planned sequence of a bioseparation scheme depending on the sample composition with respect to ionic strength and pH. The technique is amenable for laboratory scale as well as large scale separation of biological products. A few applications of IEC in the purification of proteins, isoenzymes, immunoglobulins, polynucleotides and nucleic acids are listed in this section.

IEC with a strong anion exchanger (Mono Q) has been used to recover 89% of creatine kinase from a partially purified preparation of chicken breast muscle by gradient elution of increasing ionic strength. Isoenzymes (isoforms of an enzyme) have almost the same molecular weight and hence cannot be separated by gel filtration. IEC has been used for their separation on the basis of small differences in their charge characteristics due to altered amino acid composition. Thus N-acetyl-β-D-glycosaminidase, an isoenzyme referred to as Intermediate 1 Form in the diagnosis of lymphoblastic leukaemia has been resolved into a number of component isoenzymes by high resolution IEC. Other select examples of applications include the separation and purification of monoclonal IgG_{2b} from rat cell culture, plasmids from bacterial cultures, restriction fragments of DNA and individual nucleotides, blood products such as albumin and IgG and products of rDNA technology such as growth factors.

11.6 CHROMATOFOCUSING

Chromatofocusing involves the separation of proteins by isocratically formed pH gradient on an ion exchange column. The technique has the favourable attributes of both ion exchange chromatography and isoelectric focusing. In fact, it is possible to separate proteins with similar net surface charge but differing in surface charge distribution (these are not well separated by isoelectric focusing). Another advantage of chromatofocusing is that the pH gradient is formed with a single eluent isocratically without the use of any gradient formers.

11.6.1 Principle

The basic principle of separation depends on the surface charge of protein, which differs from protein to protein. Most of the proteins have charged groups consisting of titratable functional groups on their surface and exhibit considerable heterogeneity in the distribution of these surface charges. Even though a majority of the proteins have isoelectric points in the pH range 5–8, the net charge distribution varies from protein to protein due to the presence of individual reactive groups and differences in molecular weights and subunit compositions. The surface charge on the protein can be influenced by a variety of physical and chemical factors. These include change in temperature, type and concentration of counter ions, use of water activity modifiers such as urea or ethylene glycol and the presence of metal ions, specific ligands or detergents. Proteins have a net negative charge in buffers with pH values above their isoelectric points and hence interact with positively charged anion exchange resin. Similarly, proteins will have a net positive charge in buffers with pH values lower than their isoelectric points and will not interact with anion exchange resin but will interact with cation exchange resin. A gradual change in pH of the medium will bring about a continuous change in the charge distribution on the protein surface and hence the interactions of the protein with the anion exchange resin. Proteins applied to the anionic stationary phase equilibrated at a high pH (by an equilibrating buffer) bind to the column and are eluted by a buffer (focusing buffer) of pH lower than the isoelectric point of the protein of interest. A descending pH gradient develops on the column with a steady drop of pH resulting in the elution of proteins close to their isoelectric points. It is possible to separate proteins differing in their isoelectric points by as little as 0.05 units. In the case of cationic stationary phase, an ascending pH gradient develops as mobile phase buffer percolates through the column.

The development of a pH gradient on the column internally may be explained on the basis of the equation derived by Sluyterman et al.

$$pH_f = \frac{\beta_s pH_s + \beta_m pH_m}{\beta_s + \beta_m} \tag{11.1}$$

where β_s and β_m are the buffering capacities of the stationary phase and the mobile phase respectively and pH_s and pH_m are the initial pH's of the stationary and the mobile phases and pH_f is the final pH on mixing the two phases. Assuming that β_s is equal to β_m and the buffering capacities to be additive in a simple case, the final pH (pH_f) will be a simple average of the pH_s and pH_m. The slope of the pH gradient in such a case is linear and depends on the number of column sections for a given amount of the mobile phase or focusing buffer. However, if the buffer capacity of the buffer is greater than that of the stationary phase, the slope of the pH gradient will be greater. The linearity of the slope of the pH gradient may be affected due to electrostatic or hydrophobic interactions between the constituents of the two phases and thereby the even functional buffering capacity is also altered.

11.6.2 Practice of Chromatofocusing

An ion exchange stationary phase with a strong and even buffering capacity in conjunction with a mobile phase buffer of maximum buffering power is used to separate a mixture of proteins by pH gradient elution. In general, low ionic strength of the buffer is preferred to maximize protein-anion exchange resin interaction. The stationary phase should have high porosity and rigidity. It should have a percentage of permanent or fixed (i.e. non-titratable) charges in addition to titratable charges to facilitate higher loading capacity. The permanent charges on an anion exchange resin arise due to the presence of quaternary amines while on a cation exchange resin, sulphonic acid groups function as fixed charges. The stationary phase should not have non-specific adsorption for the solutes. The buffering capacity of the stationary phase may be determined by titrating with a standard solution of a base at high ionic strength. The buffering capacity of the stationary phase is calculated from the change in pH observed for incremental addition of the base, as dB/dpH. High capacity weak anion exchange resins with a broad buffering capacity (titration range) of pH 5–10 are available commercially for use in chromatofocusing as stationary phase. Examples include Sepharose based tertiary and quaternary amine immobilized stationary phases with wide pH range stability such as Polybuffer Exchanger 94 (pH 3–12), Polybuffer Exchanger 118 (pH 3–12) and Mono P (pH 2–12) and sililca based polyethyleneimine immobilized stationary phases SynChroPak AX-300, AX-500 and AX-1000.

The chosen anion exchange stationary phase is packed into the column and equilibrated with buffer. The buffering capacity of the stationary phase in general depends on the amount of buffering groups available and the column volume. This, in turn, determines the concentration and/or volume of the mobile phase buffer required to generate a pH gradient of defined slope. It is preferable to have the smallest possible column volume (usually 15–25 cm long narrow bore columns containing 2–4 ml of stationary phase) consistent with sample load. Large column volumes require more time for regeneration. The pH gradient becomes less steep as it travels through the column, but beyond certain lengths, the gradient does not change much.

The mobile phase includes the chromatofocusing column equilibration buffer and the focusing buffer used to generate the internal pH gradient. Polymeric buffers called ampholytes have both positive and negative charges and very high buffering power but contribute very little to the overall ionic strength. Hence they can be used at high concentrations to control the pH closely. The buffering capacity of the mobile phase buffer may also be determined by titration with base or acid as in the case of the stationary phase. Commercially available polymeric buffers include Pharmacia Polybuffer 96 (titrates between pH 9 and 6) and Polybuffer 74 (titrates between pH 7 and 4).

The sample components should be completely soluble in the column equilibration buffer. Hence insoluble or immiscible components may be removed by centrifugation at high speeds (10,000–15,000 g for 10–15 min.) or by filtration through inert, porous membranes (0.22 or 0.45 μm). Proteins which are too difficult to solubilize may be treated with 1 to 6 M urea in the sample preparation buffer at 4°C or with low concentrations of detergents such as Triton X-100 or Tween 80 without compromising the biological activity of the solute protein. The sample is applied to the column and equilibrated.

11.6.3 Applications

A relatively fast separation (30 minutes) of human very-low-density lipoproteins (VLDL) was achieved by chromatofocusing the sample protein mixture obtained from the serum of hypertriglyceridemic patients. The VLDL apolipoproteins were dissolved in 0.025 M bistris (pH 6.3) chromatofocusing buffer containing 6 M urea and applied to Mono P column and chromatofocused using a descending pH gradient with Polybuffer 74. Seven peaks were resolved and collected for further identification by isoelectric focusing and SDS PAGE.

Chromatofocusing has been used both for analytical and preparative separation of molecular heterogeneous forms of ferredoxin-NADP oxidoreductase from nitrogen fixing cyanobacterium. The method has been used to analyse surface charge heterogeneity and purify various steroid receptor protein isoforms for the female sex hormone estradiol.

Questions

1. What are ion exchangers? Classify them.
2. Discuss the principle of separation of charged species by IEC.
3. Explain the terms 'total capacity', 'available capacity' and 'dynamic capacity' with respect to ion exchangers. How are they determined?
4. Write a note on the different operating modes of ion-exchange separation.
5. Describe the practice of IEC.
6. Give an account of the principle and practice of chromatofocusing.

CHAPTER

12

Reversed Phase and Hydrophobic Interaction Chromatography

12.1 HYDROPHOBIC INTERACTION

Most proteins in their native biologically active condition have hydrophobic regions on the surface. These hydrophobic regions play a predominant role in protein-protein interactions, orientation of protein molecules in the membranes as well as in the transduction of biological signals across the membranes. The interaction between hydrophobic regions at the molecular level is responsible for a variety of activities of biomolecules. The activities include the folding of globular proteins, association of protein subunits, self-association of phospholipids and other lipids to form membrane bilayers and binding of membrane proteins, binding of substrates to enzymes during catalytic reactions, regulation and transport of molecules across membrane surfaces in biological systems.

The thermodynamic aspects of hydrophobic interaction may be explained on the basis of entropy. Dissolution of a non-polar solute in water is a thermodynamically unfavourable process mainly due to a large negative change in entropy ($-\Delta S$) because of the development of an ordered structure of water molecules around the solute molecule. The change in enthalphy (ΔH), is negative as many new hydrogen bonds are formed between water molecules as an ordered structure develops. On the other hand, in aqueous salt solutions of reasonably high concentration, aggregation of the non-polar solute molecules by hydrophobic interaction, as in salting-out of proteins, is a thermodynamically favourable process. This is due to the positive change in entropy ($+\Delta S$) as the water molecules of the ordered structure around the solute molecule are released into the bulk phase.

At a given temperature, hydrophobic interaction between proteins in aqueous solution is influenced by two factors, namely, (i) the type and concentration of the salt used and (ii) presence of additives, which change the polarity of the solvent. The influence of different salts on hydrophobic interaction is predicted by Hofmeister (lyotropic) series as given below for anions and cations.

Anions

Sulphate > Chloride > Bromide > Nitrate > Perchlorate > Iodide > Thiocyanate

Cations

Magnesium > Lithium > Sodium > Potassium > Ammonium

Salts containing anions and cations at the left hand end of the series promote hydrophobic interactions better while anions to the right hand end of the series are called chaotropic. The effectiveness of different salts in promoting hydrophobic interaction is related to their contribution to the surface tension of the solution. Salts, which increase the surface tension of a liquid, promote hydrophobic interaction. Factors other than surface tension also influence hydrophobic interaction. These include protein hydration and specific interaction between the protein and salt ions. Additives such as ethylene glycol change the solvent polarity by changing the structure of water towards a structure resembling an organic solvent and thereby influence the hydrophobic interaction.

The effect of the hydrophobic interaction is exploited in the separation, concentration and purification of proteins. Salting out of proteins and organic solvent mediated precipitation of proteins are well-practiced methodologies in the downstream processing of proteins. Two chromatographic techniques, namely reversed phase chromatography (RPC) and hydrophobic interaction chromatography (HIC) are essentially based on exploiting the hydrophobic interaction for the purification of proteins.

12.2 REVERSED PHASE CHROMATOGRAPHY AND HYDROPHOBIC INTERACTION CHROMATOGRAPHY

The techniques are similar in several aspects. The chromatographic gel matrix in both RPC and HIC contain hydrophobic ligands covalently bound to the gel, but in RPC, the density of ligands is much higher compared to that in HIC. Though both the techniques use hydrophobic interaction for separation, the mechanism at the molecular level is different. In RPC the stationary phase may be considered as a continuous hydrophobic phase and the interaction between the stationary phase and the solute molecules is relatively strong. Hence only small molecules and peptides can be separated on RPC columns whereas globular proteins often denature when applied to RPC columns. In addition, RPC requires drastic conditions for elution such as a gradient of organic solvents. As against this, in HIC the ligands interact individually with solute molecules in relatively a mild manner.

12.3 BASIC THEORY OF RETENTION IN RPC AND HIC

The mechanistic aspects of the specific retention of a solute on RPC or HIC column may be explained on the basis of thermodynamic consideration of the energetics of retention. The free energy change for the chromatographic retention ΔG^0, is given in terms of the partition coefficient K (an equilibrium constant), gas constant R and absolute temperature T.

$$-\Delta G^0 = RT \ln K \tag{12.1}$$

Since the partition coefficient is related to the capacity factor as $k' = (K/\beta)$, where β is the phase ratio (ratio of the volumes of the stationary and the mobile phases, V_s/V_m), the free energy change for the retention will vary with k'.

According to solvophobic theory, various factors contribute to the free energy change for retention of the solute (protein) on the hydrophobic stationary phase. These include factors which promote binding of the solute to the stationary phase (and hence retention) as well as factors which oppose binding (and hence decrease retention). Factors which promote binding include the entropy change arising from the change in the free volume (reduction in the free volume), van der Waals and hydrophobic interactions while factors, which oppose binding, include electrostatic and van der Waals' interactions between the protein and the solvent. According to this theory, the retention factor decreases with decreasing surface tension of the mobile phase i.e. by increasing the organic solvent concentration in the eluent. Thus proteins and peptides are strongly retained on RPC columns when neat water (solvent with the highest surface tension) is the mobile phase. In contrast, in a HIC column where the hydrophobic character of the stationary phase is much weaker compared to a RPC column, retention factor increases with increasing salt concentration (surface tension increases with increasing salt concentration in aqueous solutions). The retention factor decreases with decreasing salt concentration of the eluent and neat water is a strong eluent.

Experimentally the retention of a solute in the RPC column (or HIC column) is conveniently manipulated by varying the capacity factor k' as it is strongly dependent on the composition of the mobile phase. As already mentioned in Chapter 9, for optimal performance of a column, k' should be in the range 2 to 5, but for most cases the range is extended to $1 < k' < 20$. The value of k' can be modified by changing the polarity (or solvent power) of the mobile phase. Solvents, which interact strongly with solutes, are called strong or polar solvents. The polarity of solvents is quantitatively expressed in terms of the polarity index, P' based on the solubility measurements for a solute. The polarity index of various solvents varies between –2 for highly non-polar fluoroalkanes to +10.2 for the highly polar water. Other solvents include cyclohexane ($P' = 0.04$), *n*-hexane (0.1), carbon tetrachloride (1.6), toluene (2.4), diethyl ether (2.8), tetrahydrofuran (4.0), chloroform (4.1), ethanol (4.3), dioxane (4.8), methanol (5.1), acetonitrile (5.8), nitromethane (6.0) and ethylene glycol (6.9). It is possible to arrive at any required polarity index of the mobile phase between these limits by simply mixing two appropriate solvents. The polarity index of a binary mobile phase (P'_m) is given by polarity indices of the two chosen solvents and their volume fractions (ϕ_1 and ϕ_2) as given by Eq. (12.2).

$$P'_m = \phi_1 P'_1 + \phi_2 P'_2 \tag{12.2}$$

A two unit change in the polarity index brings about a tenfold change in the capacity factor values because on an RPC column, the ratio of the capacity factors of the solute with the two pure solvents is related to the polarity index as given by Eq. (12.3).

$$\frac{k'_2}{k'_1} = 10^{(P'_2 - P'_1)/2} \tag{12.3}$$

12.4 REVERSED PHASE CHROMATOGRAPHY

The well-established normal or polar phase adsorption chromatography uses polar stationary phase such as water or triethyleneglycol supported on silica or alumina particles and a non-polar mobile phase such as hexane or iso-propyl ether. Historically the term reversed phase chromatography (RPC) was coined to describe the use of a non-polar stationary phase in conjunction with a polar mobile phase. In normal phase chromatography, the solutes elute with increasing order of polarity.

In RPC, the initial mobile phase binding conditions are primarily aqueous with a high degree of ordered structure of water surrounding the solute as well as the immobilized ligand of the stationary phase. As the solute binds to the immobilized hydrophobic ligand, the degree of ordered structure of water decreases as water molecules are released into the bulk mobile phase (positive entropy effect). The solutes elute with decreasing order of polarity as they migrate in decreasing order of net charge, extent of ionization and hydrogen bonding capabilities. Separation of polypeptides and proteins occurs due to weak van der Waals type interactions between the solute and the stationary phase of RPC. Greater selectivity is obtained depending on the protein structure (tertiary and quaternary structure) and non-polar hydrophobic interactions.

12.4.1 Practice of Reversed Phase Chromatography

Stationary phase. The stationary phases mostly use macroporous silica gel particles (5–10 μm dia.) of high rigidity and mechanical strength as the support material. The surface of the silica particles is covered completely with covalently bound non-polar functional groups. Since silica has the disadvantage of dissolving in alkaline solutions which restricts the use of the silica based stationary phases between pH limits of 2–8, polymeric supports and alumina supports of high stability in the pH range 1–13 have also come into vogue. The particle size, shape and pore structure of the support have influence on the column efficiency. In general small sized spherical particles having pore sizes in the 100–150 Å are more suitable for smaller peptides while for larger proteins pores larger than 300 Å are required. Micropellicular stationary phases having a fluid impervious core based on silica as well as polystyrene based supports have been developed for use in separation of proteins by RPC.

The actual stationary phase surface covering the silica particles in RPC involves non-polar functional groups such as *n*-alkyl chains of octadecyl, octyl or butyl groups and phenyl groups. Other non-polar functional groups such as cyano, diphenyl or fuorocarbon are also used. The non-polar phase is bound covalently to the silica particle surface by treating the surface silanol groups with appropriate silanizing agents. The reaction may be represented as shown below.

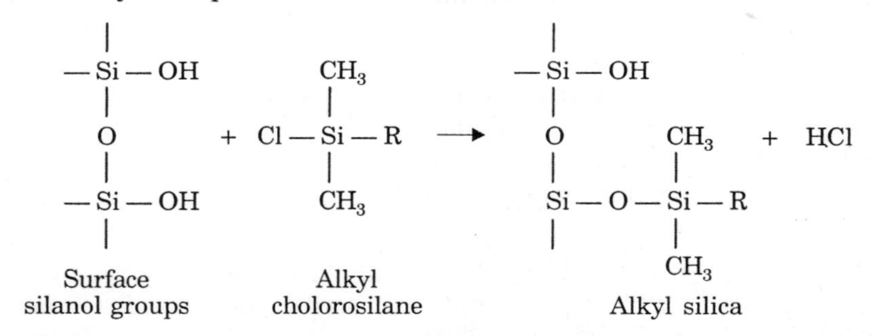

Figure 12.1 Covalent binding of non-polar functional groups on silica surface.

The nature and density of the non-polar groups at the surface affect the retention, the loading capacity and selectivity of the column. The surface silanol groups cannot be completely converted by silanization and the presence of the residual silanol groups render the stationary phase heterogeneous. Both hydrophobic (solvophobic) as well as silanophilic interactions are exhibited by such a stationary phase surface towards the protein solute. Electrostatic interactions with free residual silanol groups can result in peak asymmetry and irreversible adsorption of proteins. The silanophilic interactions are reduced by endcapping the silanol groups by treatment with trimethylchlorosilane so that the surface is homogeneously non-polar.

Mobile phase. The mobile phase in RPC is usually a water-organic solvent mixture containing a buffer component, an organic modifier such as methanol, isopropanol, acetonitrile or tetrahydrofuran and often an ion pairing agent to affect the selectivity. Solute retention decreases with increasing concentration of the organic modifier (usually expressed in terms of per cent v/v). The organic modifier affects the gross properties of the mobile phase such as surface tension, viscosity, dielectric constant etc. The surface tension of the mobile phase is important in determining the magnitude of retention while viscosity determines the pressure drop across a given column and also the efficiency of separation by its effect on diffusion rates of the solutes in the sample. The organic modifier may also interact with the protein solutes specifically inducing structural changes and denaturation. For example, low concentrations of alcohols (< 0.1% v/v) stabilize chymotrypsinogen while higher concentrations induce reversible or irreversible structural deformation.

The eluotropic strength of the mobile phase increases with increasing content of the organic modifier. The eluting strength of the mobile phase for eluting small molecules from an RPC column follows the order methanol < acetonitrile < ethanol ~ acetone ~ dioxane < isopropanol < tetrahydrofuran, but no such eluotropic series is available for proteins. Acetonitrile and methanol find extensive use as organic modifiers in the RPC of peptides and proteins. High adsorption isotherms are observed with binary mobile phase which may lead to denaturation of proteins while a ternary mobile phase containing two organic solvents can prevent denaturation of proteins as the composition tends to decouple eluting strength from the denaturing strength. In addition, a ternary mobile phase provides improved selectivity and hydrophobic proteins may be eluted out with a mobile phase of higher water content.

Peptide and protein samples are generally separated by gradient elution with increasing organic solvent content in the mobile phase. Proteins bind to the hydrophobic stationary phase rather strongly and hence isocratic elution is not helpful. The pH of the mobile phase has a profound influence on retention, selectivity of separation as well as peak shape. RPC separations are best carried out in the pH range 2–4. At higher pH values and particularly between pH 5 and 7, the chromatographic peaks are generally broader and highly asymmetric. Hence, most peptides and proteins are best separated with acidic eluents containing trifluoroacetic or phosphoric acid at about pH 3–3.5. Trifluroacetic acid (0.1% v/v) increases the retention time of proteins and peptides while phosphoric acid decreases the retention time. Trifluoroacetic acid has advantages in that it is optically transparent above 210 nm and hence permits detection of peptide bonds, has a good solubilizing effect on proteins and peptides and being highly volatile, can be easily removed from samples after separation. At such low pH values, the residual free silanol groups of the stationary phase as well as the carboxyl and amino groups of the proteins and peptides are protonated and do not exhibit electrostatic interactions. The amino groups of the solutes form ion pairs with trifluoroacetic acid and give sharp elution peaks. Addition of chaotropic salts to the mobile phase results in weaker retention and also increases peak sharpness.

Hetaeric chromatography is a technique in which a complexing ion (hetaeron) is added to the mobile phase to influence the selectivity of RPC by secondary equilibria. Many non-polar and polar hetaerons are known and they may be classified into two groups; the bulky, surface active ions of molecular weight ~ 200 such as alkyl sulphates and quaternary alkylammonium salts; and the smaller polar ions with low adsorption isotherms for the hydrocarbonaceous stationary phase such as phosphate, perfluoroacetate or bicarbonate salts. These co- or counter hetaerons form ion pairs with proteins and peptides and hence their addition to mobile phase is useful in altering retention character-istics. Tri- or tetraalkylammonium phosphate buffer and ammonium bicarbonate are widely used in RPC.

Chromatographic technique. The steps involved in the use of RPC are (i) column selection, conditioning and equilibration (ii) sample preparation and application (iii) washing (iv) elution and (v) column regeneration and storage.

(i) *Column selection, conditioning and equilibration:* Packed columns of stationary phases for use in RPC mode of HPLC are commercially available. Examples include the most widely used RP-18 or C-18 octadecyl column with 18 carbon hydrocarbon chain, RP-8 or C-8 octyl column with 8 carbon hydrocarbon chain as the stationary phase. The choice of the reversed phase matrix is based on requirements of a specific application. These include the type of sample components, molecular weight or size and hydrophobicities of the sample components. In addition, the scale of operation and mobile phase conditions also need to be considered. The nature of the sample components (e.g. peptides or large molecular weight proteins or oligonucleotides etc.), their source (e.g. native sources or synthetic origin or from enzyme digests), their stability and biological activity are of importance in selecting the RPC matrix. The accessibility of the sample components to the immobilized ligands of the RPC gel depend on the molecular size of the sample components and pore size of the gel. Generally, pore sizes of 30 nm or larger are used for the separation of proteins and pore size less than 30 nm are useful for peptides and oligonucleotides. The greater the hydrophobicity of the sample components, the less hydrophobic the RPC phase should be; the hydrophobicity of the RPC phase increasing with the length of the hydrocarbon chain (C-18 > C-8 > C-2). RPC columns need to be conditioned for first time use or after long term storage or when mobile phase conditions are changed significantly. Usually the column is conditioned and equilibrated by using the same mobile phase to be used in the chromatographic separation, till a stable base line is obtained prior to sample introduction.

(ii) *Sample preparation and application:* It is preferable to dissolve the sample in the initial mobile phase. Increased solubility and stability of the sample may be achieved by adding salt or acids such as formic or acetic acid. The sample should be free from turbidity and hence needs to be centrifuged or filtered through a microfilter. The sample volume should be small compared to the column volume. The sample is applied at a flow rate where optimum binding will occur.

(iii) *Washing:* Once the sample is applied, the column is washed with the initial mobile phase to remove any unbound and occluded sample components.

(iv) *Elution:* Elution of the adsorbed solute components is carried out by adjusting the polarity of the mobile phase composition so that the bound components desorb sequentially. The polarity of the mobile phase is usually decreased (increasing hydrophobicity) by gradually increasing the percentage of the organic modifier from an initial level of 0% to a final mobile phase of maximum content of 100% (or less) of organic modifier. The solute components elute out according to their individual hydrophobicities.

(v) *Column regeneration and storage:* After elution of the components, any residual components not desorbed during the elution step is washed out by changing the mobile phase to near 100% of organic modifier. The column is then re-equilibrated with the initial mobile phase. The RPC gel may be stored in 20% ethanol or pure methanol.

12.4.2 Applications

Chromatotopography is a technique to study the conformational changes induced by either the mobile phase or the stationary phase in the three-dimensional structure of the proteins provided the intermediates have half-lives at least 10 times the time required for the protein to traverse a theoretical plate. Peptides and proteins of molecular weight smaller than 65,000 Da can be handled by RPC. Protein samples tend to undergo unfolding and denaturation in RPC because of the harsh conditions employed. The various factors which cause loss of biological activity of proteins include (i) solvent induced denaturation wherein the mobile phase composition brings about conformational changes independent of the stationary phase particularly when the organic modifier concentration exceeds 15% v/v as in the case of catalase. Propanol was found to induce helical conformation and the native forms of the proteins could be restored by the removal of the organic modifier. (ii) Stationary phase induced denaturation independent of the mobile phase as in the case of pepsin and β-lactoglobulin on alkylsilica. The stationary phase catalyzes the unfolding of the protein chain and depending on the relative rates of unfolding and refolding as well as the separation kinetics unusual peak shapes and multiple peaks may be observed. For example, RPC of papain yields two peaks, the first one due to the native protein. The size of the second peak due to the denatured papain increases with increasing residence time of the protein on the column. (iii) Conformational changes and denaturation also occur due to the combined effects of the mobile phase and the stationary phase.

Purification and analysis of proteins and peptides: Purification of insulin and insulin related peptides and insulin-like growth factor are some of the examples of the use of RPC. In the purification of γ-interferon by RPC, two peaks corresponding to two proteins, both exhibiting antiviral activity but differing in the degree of glycosylation were obtained. RPC has been used in the purification of interleukin-2 (IL-2) produced by recombinant *E. coli*. The IL-2—because of its greater hydrophobicity—has longer retention time compared to other protein impurities. The resolving power of RPC is demonstrated in the separation and purification of ribosomal proteins. Eucaryotic and procaryotic ribosomes are composed of one large subunit and a small subunit containing a large number of different proteins. A total of 32 different proteins of the large subunit and 21 different proteins of the small subunit of *E. coli* ribosome respectively have been separated by RPC and reconstituted into functional ribosomal proteins after chromatographic separation.

Separation of diastereomeric peptides, membrane proteins: Separation of diastereomeric peptides containing more than 10 amino acid residues is possible only by RPC. Separation is achieved because of the differences in their hydrophobic binding surface as shown in the case of bombesin from D-[Met_{14}]-bombesin on octadecyl-silica column. Membrane proteins have both highly hydrophilic glycan moieties and highly lipophilic polypeptide chains and are strongly retained on RPC columns. Hence their elution requires specific mobile phase composition such as 60% formic acid with increasing concentration of isopropanol. For example, virus envelope proteins have been separated with a gradient of ethanol-butanol in 12 mM HCl.

Protein sequencing and structural studies (peptide mapping): RP-HPLC is widely used in protein sequencing and structural studies and peptide mapping. For example, tryptic digest of human thyroglobulin (650 kDa) yields 470 peptide fragments which can be analysed on short columns of micropellicular C-18 silica in a short time.

Quality control of protein products: RPC is routinely used for measuring protein purity and for the characterization of protein products in quality control for proteins produced by recombinant DNA technology. For example, desamido and sulphoxide derivatives of highly purified human growth hormone have been separated from the parent compound by isocratic RPC for analysis and characterization purposes.

Trace enrichment and concentration of proteins: The fact that RPC *n*-alkylsilica columns exhibit large adsorption isotherms for polypeptides and proteins may be exploited for enrichment and concentration of trace quantities of proteins from dilute solutions without affecting the loading capacity of conventional RPC columns. For example, loading up to 50 mg of ribonuclease from a solution of about 500 ml could be achieved without affecting a C-18 analytical (25 × 0.4 cm) column.

12.5 HYDROPHOBIC INTERACTION CHROMATOGRAPHY

The hydrophobic interaction between proteins in solution and hydrophobic ligands covalently bound to a gel matrix of a HIC column is schematically shown in Figure 12.2. The changes in enthalpy and entropy for hydrophobic interaction are influenced by temperature and in general hydrophobic interaction increases with an increase in temperature at least in the temperature range of interest to HIC.

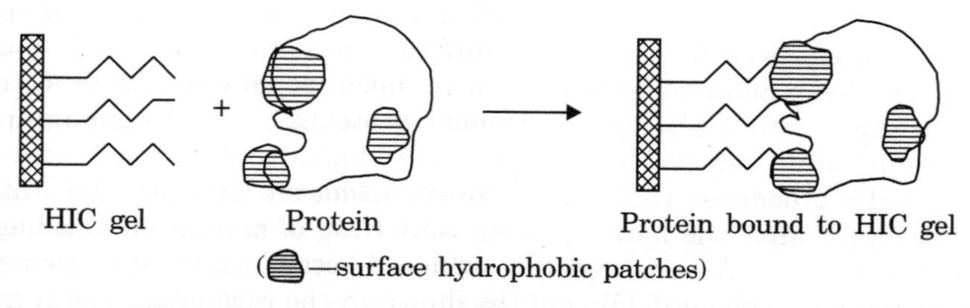

Figure 12.2 Schematic representation of hydrophobic interaction between immobilized ligand and a protein in solution.

A multipoint attachment between the HIC gel and the protein seems to be necessary for effective use of this technique as shown in Figure 12.2. In general, in a globular protein the hydrophilic amino acids are exposed at the surface with the majority of the hydrophobic amino acids buried in the interior. A smaller percentage of hydrophobic amino acids however appear on the surface as patches. The availability of such hydrophobic patches as well as their distribution on the surface of a protein is of importance in HIC. For example, when α-chymotrypsinogen is activated by cleavage of four peptide bonds, the hydrophobic amino acids ile-16 and val-17 are buried while the more hydrophilic amino acids met-192, gly-193 and arg-145 are exposed at the surface of α-chymotrypsin making it more hydrophilic than its zymogen. The enzyme elutes earlier from a HIC column compared to its zymogen indicating the importance of availability of hydrophobic patches on the surface of the protein molecule. The distribution of the hydrophobic patches on the surface is also important. For example, the molecular weights of lysozyme, myoglobin and ribonuclease are similar and may be expected to have similar number of ligand attachments and similar binding to a HIC gel. However lysozyme with a smaller per cent of hydrophobicity (41, per cent of hydrophobicity compared to 48 and 46 per cent for myoglobin and ribonuclease respectively) binds to HIC gel more strongly compared to myoglobin indicating that the distribution of the hydrophobic patches on the surface of the protein is also important in deciding the degree of binding to a HIC gel. Kinetic studies on the binding of phosphorylase-*b* to butyl Sepharose gel have indicated that the binding of a protein to the HIC gel is a multistep reaction. The rate limiting step seems to be the conformational change or reorientation of the protein on the gel.

HIC is relatively a mild technique due to the presence of high concentration of salts, which stabilizes proteins. The recovery of solutes is quite high. In a study of model proteins of cytochrome, ribonuclease, transferrin and lysozyme on octyl-agarose HIC column the chromatographic elution patterns obtained by elution immediately after the application of the sample and after several hours of retention on the HIC column were identical. This clearly indicates the technique to be a mild one because the protein structures were not altered due to binding to a HIC column even for prolonged time.

However, more labile proteins can undergo changes in their structure in contact with a HIC gel as shown in the case of α-lactalbumin. The protein, after removal of calcium ions which stabilize the structure, showed a broadened peak with a longer residence time on a HIC column that eluted after the native protein consisting of several unfolded species of the protein. A rerun of the second fraction again gave two peaks indicating that the unfolding of the protein to be reversible. The retention of a protein on a HIC column also increases with increase in temperature due to the increase in the hydrophobic interaction as in the case of lysozyme. Structural changes in more labile proteins also enhance the retention time. Thus at higher temperatures α-lactalbumin shows a higher proportion of unfolded protein. HIC may be used as a second step of purification after salting out the most hydrophobic proteins in a given mixture as it gives a sharper separation. HIC can be followed by gel filtration to remove the salt but not ion exchange chromatography.

12.5.1 Practice of HIC

The steps involved are (i) preparation of HIC gel (ii) column packing and equilibration (iii) sample application (iv) elution and (v) regeneration of column.

Preparation of HIC gel. Most extensively used gel is based on agarose. Other matrices include dextran and cellulose, which are amenable for high salt concentrations, the technique employed in such cases being called salting-out chromatography. For HPLC mode of HIC, silica and organic polymer resin based gels of uniformly small size (~ 1.5 micrometer) are used to withstand high pressure operation. Commercial gels include phenyl-Sepharose CL 4B, octyl-Sepharose CL 4B, phenyl-Sepharose FF, alkyl-agarose and butyl-toyopearl 650. For FPLC, phenyl-Superose, alkyl-Superose and silica based Synchropak (hydroxylpropyl, methyl, benzyl, butyl or pentyl groups) are available. Different hydrophobic ligands may be coupled to the matrix by glycidyl ether coupling to an agarose gel as shown by the scheme below.

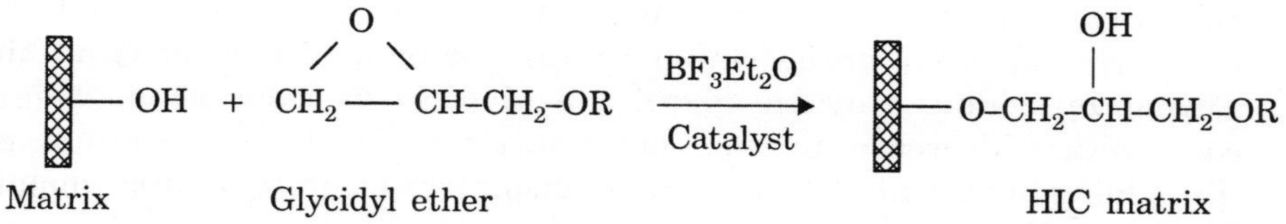

$$\text{Matrix}-OH + \underset{\text{Glycidyl ether}}{CH_2\overset{O}{—}CH\text{–}CH_2\text{–}OR} \xrightarrow[\text{Catalyst}]{BF_3Et_2O} \underset{\text{HIC matrix}}{\text{Matrix}-O\text{–}CH_2\text{–}\overset{OH}{\overset{|}{C}H}\text{–}CH_2\text{–}OR}$$

Figure 12.3 Coupling of hydrophobic ligand to matrix.

The density of ligand coupled to the matrix is important for achieving good results. Commercially available octyl- and phenyl-Sepharose gels have a ligand density of 40 micromoles/mol of gel bed corresponding to a degree of substitution of approximately 0.2 moles of hydrophobic substituent per mole of galactose.

HIC gels can be used for batch adsorption as well as large-scale chromatography. HIC gels should be preferably charge free, otherwise at low ionic strengths, proteins can interact with positively charged amino groups on HIC gels. The strength of hydrophobic interaction between the HIC gel and the sample protein increases with increasing length of the alkyl ligand coupled to the gel. Ligands containing 4 to 10 carbon atoms are suitable for separation. For membrane proteins, which are poorly soluble in buffers of high salt concentration, long chain ligands on HIC gels give better separation. HIC gels with aromatic ligands show pi-pi interaction besides hydrophobic interaction. A phenyl group has about the same hydrophobicity as a pentyl group but differs in its selectivity due to pi-pi interaction as aromatic ligands on protein surfaces can interact specifically with phenyl and other aromatic ligands on HIC gels. HIC gels with low ligand density interact less strongly and are amenable for isocratic elution than HIC gels with high ligand density.

The steps after gel preparation include column packing and equilibration, sample application, elution and column regeneration and represent the chromatographic technique.

Column packing and equilibration. The HIC gels are packed into columns and equilibrated with buffers of 0.01 to 0.05 M strength containing salt of appropriate concentration. Usually ammonium sulphate in the concentration range of 0.75 to 2.0 M or sodium chloride in the range of 1 to 4 M are used. The salt concentration should be below the concentration that precipitates out the proteins by salting-out.

Sample application. The sample solution in appropriate buffer containing the chosen salt of appropriate concentration should be clear, otherwise it should be centrifuged or filtered and the supernatant sample should be used. The pH of the buffer has a decisive influence on the adsorption of proteins to a gel. Labile protein should be chromatographed at low temperatures to prevent their denaturation.

Elution. Stepwise or gradient elution can be carried out by (i) decreasing salt concentration or (ii) changing the polarity of the solvent by adding ethylene glycol (up to 80%) or isopropanol with a concomitant decrease in the salt concentration in the column or (iii) adding detergents which work as displacers of proteins used mainly for the purification of membrane proteins.

Regeneration of HIC gels. After chromatographic run, strongly adsorbed proteins may be washed out with 6 M urea or guanidine hydrochloride. The gel is then equilibrated with starting buffer to regenerate it. If detergents have been used the gel should be washed with different alcohols. The regenerated gel can be stored at 4°C in the presence of 20% ethanol.

12.5.2 Applications

By changing the elution conditions a wide variety of separations and purifications have been achieved using HIC. A few representative applications are discussed here.

Elution with a decreasing salt concentration has been used in the purification of acid phosphatase from bovine cortical bones. The protein mixture was extracted from washed and homogenized bovine long bones (tibia). The crude extract was subjected to a four step-purification methodology involving CM-Sepharose ion-exchange chromatography, cellulose phosphate affinity chromatography, Sephacryl S-200 gel filtration and finally by HIC on phenyl-Sepharose column. The active fractions from gel filtration were pooled and adjusted to 30% ammonium sulphate saturation at 4°C and then applied to phenyl-Sepharose HIC column equilibrated with sodium acetate buffer (pH 6.5) containing ammonium sulphate at 30% saturation. The column was washed with the same equilibrating buffer and then eluted with a decreasing salt concentration from the initial 30% to 0% in sodium acetate buffer (pH 6.5). The enzyme was recovered to the extent of 92% with a 13 fold purification in the HIC step.

Elution by decreasing polarity of the eluent has been used in the purification of human pituitary prolactin. The protein extract obtained from homogenized frozen pituitary glands was subjected to a two-step purification process. The extract was first subjected to gel filtration on Sepharose CL-6B at pH 9.8 and fractions containing prolactin were pooled and chromatographed on phenyl-Sepharose column equilibrated with 0.2 M glycine-sodium hydroxide buffer (pH 9.8). Prolactin was recovered (95% recovery) by eluting out with an eluent of decreasing polarity from 0.2 M equilibrating buffer to 0.02 M concentration containing 50% v/v ethylene glycol. Further purification of prolactin was carried out by gel filtration on Sephadex G-100 superfine followed by ion-exchange chromatography on DEAE-Sepharose CL-6B.

Elution using a detergent is yet another method by which protein separation/purification can be achieved in HIC. In the case of membrane bound proteins, purification steps involve the use of more than one detergent and in such cases exchange of detergents bound to membrane proteins may be achieved by using HIC columns equilibrated with the appropriate detergent solution.

Questions

1. Explain the principle involved in reversed phase and hydrophobic interaction chromatographic techniques.
2. Distinguish between the mechanisms of reverse phase and hydrophobic interaction chromatographic techniques.
3. Describe the practice of reversed phase chromatography. What are its applications?
4. What are the factors that affect the hydrophobic interaction between proteins and the ligands?
5. Describe the practice of HIC with suitable examples.
6. Write a note on the stationary phases used in RPC.
7. How are HIC gels prepared?
8. What is chromatotopography?

CHAPTER

13

Affinity Chromatography

13.1 GENERAL FEATURES

Affinity chromatography includes a group of closely related techniques such as bioaffinity, dye-ligand affinity and immobilized metal ion affinity chromatographic techniques. In all these techniques the stationary phase matrix is specially prepared to isolate or purify biomolecules based on specific interactions between the matrix immobilized functional groups and functional groups on the surface of the biomolecules. The less frequently used covalent chromatography may also be included in this group as the matrix is specially prepared to separate biomolecules having surface exposed sulphur containing functional groups. All these chromatographic techniques are relatively mild and capable of high resolution and recovery of the desired solute component from a complex mixture.

13.2 SPECIFIC AND NON-SPECIFIC INTERACTIONS

Chromatography involves some kind of interaction between the solid phase and the components of the sample undergoing separation, purification or concentration. The nature of these interactions may be simple or complex. However, in almost all the chromatographic techniques the interactions are non-specific in that the similar components in a given mixture interact with the stationary phase in a similar manner and the separation is based mainly on the minor differences in the extent of interaction. An example of a method involving a simple and well-defined interaction is ion-exchange chromatography, which depends on the affinity of a charged species for its counter-ion. Thus, it is possible to immobilize a cation in order to selectively adsorb and hence purify anions or vice versa. However, the method is non-specific because all ions of like charge will either be adsorbed or not adsorbed. The same drawback exists for all other types of chromatographic techniques. The lack of specificity is a serious one, which affects the separation efficiency as well as recovery of the desired product. Thus extraction of an enzyme, for instance, from a natural source usually involves a succession of mild methods sequentially, which can be tedious and inefficient, resulting in diminishing recoveries. *Affinity chromatography* or *bioaffinity chromatography* overcomes this problem because of its high specificity and hence is of immense use in bioseparations.

13.3 BIOAFFINITY CHROMATOGRAPHY

The term bioaffinity chromatography is used to include different types of chromatography involving only biologically functional pairs such as enzyme-inhibitor or antigen-antibody complexes. Dye ligand chromatography involving biomimetic ligands particularly dyes, which bind active sites of functional enzymes and immobilized metal ion affinity chromatography (IMAC) are treated as pseudo-affinity chromatographic techniques.

13.3.1 Advantages of Bioaffinity Chromatography

Advantages of this method include:

- High selectivity compared to other purification techniques based on molecular size, charge, isoelectric point etc.
- Extremely good purification up to several thousand folds in a single step and recoveries greater than 90% can be expected provided conditions are carefully selected.
- Affinity chromatography has a high concentrating effect. This is particularly true when the protein of interest is a minor component of a complex mixture. This advantage has been amply demonstrated in the extraction/purification of 10 mg active transcobalamin II (vitamin B_{12} transport protein) from 40 kg of plasma in a two-step procedure using column immobilized cobalamin as ligand.
- Affinity methods can also be used to remove unwanted materials from a mixture, e.g., serum free from a particular substance can be prepared for use as standard in radio-immunology.

13.3.2 Principle of Bioaffinity Chromatography

The natural affinity displayed between biological macromolecules and complementary ligands is well known. For example, an enzyme has affinity for its substrate, cofactor or inhibitor, an antibody for the antigen which stimulated the immune response, a lectin for some sugar residue, a hormone for its carrier or receptor, a polynucleotide for its complementary strand and so on. Bioaffinity chromatography exploits this mutual biological affinity to effect separation or purification of the component from complex mixtures.

The technique of affinity chromatography exploits the formation of specific and reversible complexes between a pair of biomolecules. One of the pair is called the ligand and is usually immobilized onto a stationary phase while the other, called the counter-ligand, is adsorbed from the extract that is passing through the chromatographic column containing the immobilized ligand. Examples of biological interaction between such ligands and counter-ligands used in bioaffinity chromatography are listed in Table 13.1.

Table 13.1 Ligands and Counter-ligands in Affinity Chromatography

Ligand	*Counter-ligand*
Enzyme	Substrate, substrate analogue, cofactor or inhibitor.
Sugar	Lectin, enzyme or sugar binding proteins.
Antibody	Antigen, virus or cells.
Nucleic acid	Nucleic acid binding protein, enzyme or histone.
Lectin	Glycoproteins, polysaccharide, membrane proteins, cell surface receptor or cells.
Hormones and vitamins	Receptor or carrier proteins.

The specificity or affinity exhibited by a ligand to the counter-ligand is due to a combination of different types of interactions such as van der Waals' forces, hydrogen bonding and hydrophobic interactions. The interaction between two biomolecules, A and B, may be represented by an equilibrium as in Eq. (13.1).

$$A + B \underset{k_{-1}}{\overset{k_1}{\rightleftarrows}} AB \qquad (13.1)$$

The equilibrium constant, K (the ratio of the rate constants of the forward and reverse reactions k_1 and k_{-1} respectively) depends on the nature of interaction and the interacting biomolecules. For example, the interaction between avidin and biotin is extremely strong as reflected in the magnitude of the value of K being in the vicinity of 10^{15}. In contrast, the interactions between an enzyme and its substrate or lectin and carbohydrate are relatively more labile with K values in the range of $10^4 - 10^6$. The antigen-antibody or hormone-receptor interactions have K values in the range of 10^6 to 10^{12}. In general, if the K value is too low, affinity adsorption of the counter-ligand does not occur while if the K value is very high, the strong affinity may lead to irreversible adsorption.

In either case, separation or purification of the counter-ligand may be difficult. In such cases, change in pH, addition of salt or detergents, which affect the stability of the ligand-counter-ligand complex without destroying the active conformation of the biomolecule, may be necessary.

In practice, the ligand is usually attached covalently (immobilized) to a water-insoluble polymer stationary phase matrix or gel (such as, agarose, dextran, cellulose, polyamide or porous glass) to form a tailor-made chromatographic adsorbent material suitable to adsorb specifically the desired component(s) from a mixture. The adsorbent is packed into a column and the sample mixture is applied to the adsorbent bed facilitating the adsorption of the desired component. All other non-complimentary constituents of the mixture do not 'recognize' the ligand and pass through, unrestrained. The column is washed with a suitable buffer to remove any non-specifically adsorbed components of the sample leaving behind the desired counter-ligand bound to the stationary phase matrix. The adsorbed component may then be eluted out in pure form by using a suitable eluent or buffer. A typical separation procedure is conceptually shown in Figure 13.1. The chromatogram consists of two peaks, a first peak due to all the non-retained components of the sample and the second peak due to the elution of the desired component.

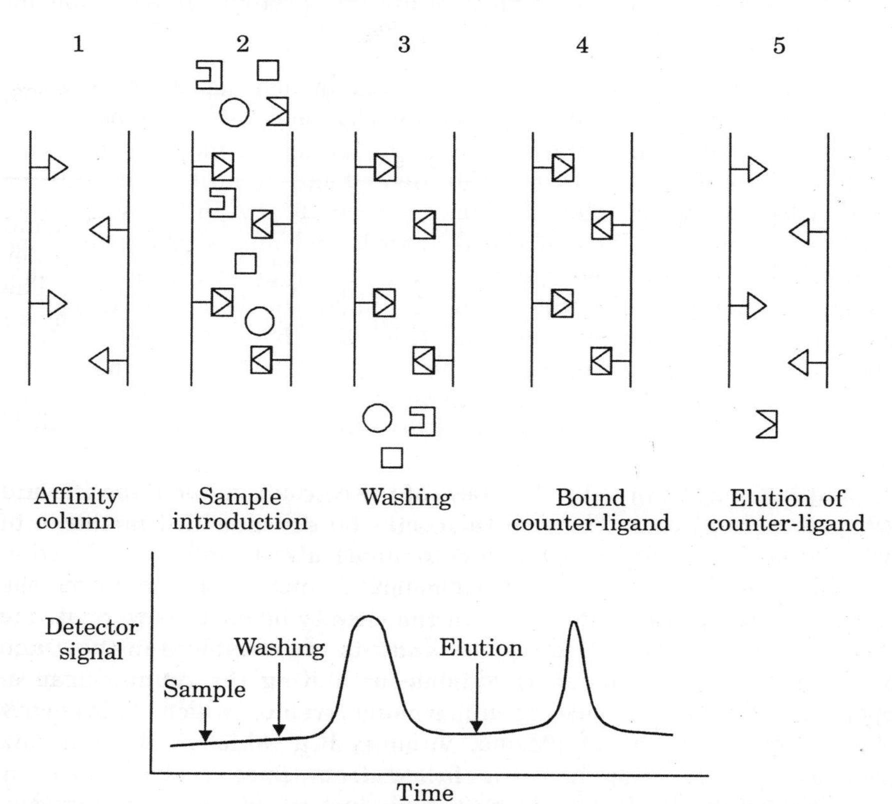

Figure 13.1 Conceptual representation of the principle of affinity chromatography and the on-line chromatogram.

13.3.3 Practice of Bioaffinity Chromatography

The steps involved in the practice of affinity chromatography include the following.

Choice of support matrix. The support matrix material should have a high surface area. It should be water-insoluble and physically rigid to withstand the operational stress. At the same time, it should be permeable and macroporous. It should contain reactive functional groups to derivatize and bind ligand molecules covalently. It should be chemically stable towards harsh conditions during derivatization or regeneration. It should have minimal non-specific adsorption characteristics. It should be reusable and preferably of low cost. Although it is difficult to identify any single matrix that can meet all the ideal requirements, the commonly used matrices are porous glass, silica, cellulose, polyacrylamide, agarose and dextran. Agarose, a natural polymer based on alternating 1,3 linked α-D-galactopyranose and 1,4 linked 3,6-anhydro-L-galactopyranose residues, has a number of features making it attractive for bioaffinity chromatography. It is available commercially in several sizes and porosities, allowing homogeneous column packing. It has no ionic groups, yet it is hydrophilic. It exhibits little affinity for proteins. It is suitable for chemical modification for attachment of ligands.

Choice of ligand. The choice of the ligand depends mainly on two factors, namely, (i) availability of chemically modifiable groups on the ligand to facilitate its attachment to the matrix while still retaining binding capability towards the counter-ligand and (ii) affinity towards the counter-ligand with K values ranging between 10^4–10^8 in free solution to ensure maximum binding of the counter-ligand and at the same time desorption of the counter-ligand under mild conditions. The functional groups commonly used for the immobilization of the ligands to the matrix include amino, carboxyl, aldehyde, thiol and hydroxyl groups. It is preferable to have 'multi-point attachment' of the ligand to the matrix rather than a 'single point attachment' to prevent any leakage of the ligand from the matrix over a period of time or during relatively harsh conditions of elution.

Ligands may be broadly classified into two groups as *monospecific* and *group specific* ligands. These in turn, may be sub-classified further into low molecular weight ligands and macromolecular ligands.

Monospecific low molecular weight ligands include steroid hormones, vitamins and enzyme inhibitors. Usually these ligands show relatively greater affinity towards their counter-ligands with K values in the range of 10^8–10^{10} requiring harsh conditions for eluting the counter-ligands. Specific examples of this group include lysine, which binds only plasminogen from blood plasma, vitamin B_{12}, which binds only its transport protein, (or intrinsic factor from pure gastric juice, or transcobalamin II from plasma) and biotin which binds avidin.

Monospecific ligands bind the counterligands strongly and hence require harsh conditions for elution which is more likely to denature the desired protein.

Group specific, low molecular weight ligands include a wide variety of enzyme cofactors and their analogues. Specific examples include 5′-AMP which binds NAD^+ dependent dehydrogenases and ATP dependent kinases; NAD^+ which binds NAD^+ dependent dehydrogenases and ATP which binds ATP dependent kinases.

Monospecific macromolecular ligands are involved in specific interactions among proteins, multi-enzyme systems, hormone-receptor protein interactions and immunosorption. For example, gelatin specifically binds fibronectin, thrombin (or heparin- a polysaccharide) binds antithrombin and transferrin binds its receptor protein.

Both antigens and antibodies may be used as affinity ligands in immunosorption. Immunosorbents are useful in the purification of proteins, peptides and solubilized membrane proteins, viruses and even whole cells. Immunoadsorbents based on polyclonal antibodies as well as the highly specific monoclonal antibodies have been widely used in the isolation and purification of antigens. *Immunosubtraction* is an innovative procedure involving the use of immunosorbent with low antibody loading for removing low concentrations of known impurities from protein products.

Group specific macromolecular ligands include lectins such as con A and lentil used in the isolation of glycoproteins, staphylococcal protein A and streptococcal protein G used for purification of IgG and calmodulin used for the isolation of calcium dependent enzymes. Sulphated heparin finds use in the purification of coagulation proteins (such as, antithrombin III, Factor VII, Factor IX, Factor XI, Factor XII, XIIa, thrombin), plasma proteins (such as, properdin, Complement C1, C2, C3 and C4, complement factor B, Beta 1H, fibronectin, C-reactive protein, Ge globulin) enzymes (e.g., lipoprotein lipase, hepatic triglyceride lipase, restriction and endonucleases, RNA and DNA polymerases), lipoproteins (such as, VLDL, LDL, VLDL apoprotein) steroid receptor, oestrogen receptor, androgen receptor and virus surface antigens (such as, hepatitis B surface antigen, SV 40 tumor antigen).

Spacer arm. Adsorbents prepared by coupling small ligands directly to the matrix often exhibit low binding capacity due to steric constraint which hinders the interaction between immobilized ligand and the active site of the counter-ligand as shown in Figure 13.2a. A spacer arm eliminates such a hindrance as shown schematically in Figure 13.2b.

The use of spacer arm for efficient binding has been amply demonstrated in the case of egg white protein avidin binding to immobilized biotin. Though the avidin-biotin interaction has a large K value (10^{15}) indicating very high affinity between the two molecules, avidin is only slightly retarded on biotinyl cellulose column. In contrast,

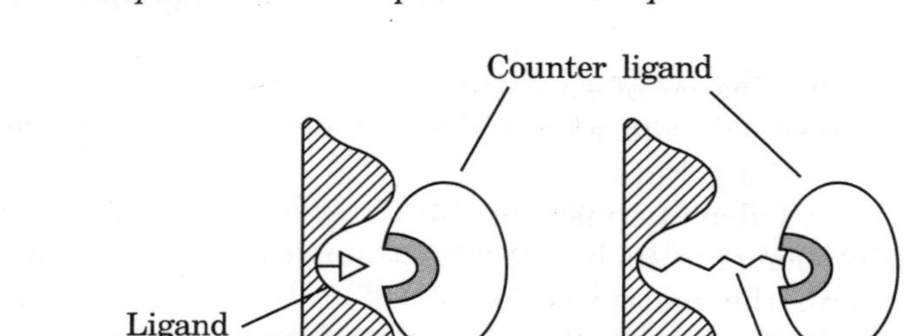

Figure 13.2 Schematic diagram of the concept of spacer arm.

avidin is strongly held on biotinyl-L-lysyl-agarose, which has a six carbon atom spacer (lysyl group) between the matrix (agarose) and the ligand (biotin) and requires a harsh eluting agent such as guanidine hydrochloride at pH 1.5.

Commonly used spacer arms are mostly linear aliphatic hydrocarbons with terminal functional groups to facilitate the binding of the spacer molecule to the matrix as well as the ligand. These include 6-amino-hexanoic acid ($H_2N(CH_2)_6COOH$), hexamethylenediamine ($H_2N(CH_2)_6NH_2$) and 1,7-diamino-4-aza-heptane(3,3-diamino-dipropylamine). Spacer molecules with less than six carbon atom chain have been found to be less effective. The spacer arm may be attached to the matrix before coupling the ligand by a procedure similar to the immobilization of the ligand to the matrix. However, excess of spacer groups can lead to undesirable hydrophobic or ionic modification of the matrix. Alternatively, a ligand may be synthesized by incorporating the spacer so as to be suitable for direct attachment to the matrix, as in the case of bitonyl-L-lysine.

Coupling of the ligand. The steps involved in the coupling of the chosen ligand to the chosen matrix include (i) chemical activation of the matrix (ii) immobilization of the ligand via the chosen functional group and (iii) blocking or deactivating the residual active groups. Most commonly the chemical activation involves the –OH groups of the support. Immobilization of the chosen ligand involves the –OH, –COOH, $-NH_2$ or –SH groups of the ligands. For maximum efficiency of the affinity adsorbent, the coupling of the ligand should give a matrix with an optimum density of ligands and minimal leakage of the immobilized ligand. The different immobilization methods are discussed later in this chapter. The first two steps of activation of the matrix and coupling of the ligand leave behind residual activated groups on the matrix, which may lead to non-specific adsorption by such groups. Hence the residual activated groups are deactivated by reacting with excess of suitable low molecular weight substances such as ethanolamine or mercaptoethanol under the same conditions used for coupling the ligand.

The ligand density (degree of substitution as micromoles or mg of ligand per ml of affinity gel) and adsorption characteristics (specifically adsorption capacity) of the gel are usually determined prior to using the affinity gel. The ligand density may be determined quantitatively by elemental analysis of elements such as N, S or P of the ligand which are normally not found in the matrix or by absorption spectroscopy at a wavelength specific to the ligand. The adsorption capacity of a gel is defined as mg or micromoles of counterligand that can be adsorbed per ml of the affinity gel. The maximum theoretical binding capacity of a 60 kD protein on a beaded agarose of 90 micron average diameter is approximately 80 mg/ml. However, in practice it is not possible to achieve this capacity. The static capacity of a gel depends on the density of the ligand and its availability for interaction with a particular protein and is measured in batch experiment allowing sufficient time for equilibration. The ligand may be inaccessible particularly to high molecular weight proteins and hence the binding capacity will be much lower compared to the density of ligand. In contrast, the dynamic capacity of a gel is the binding capacity under operating conditions in a chromatographic column.

The affinity adsorbents may be stored for longer periods at 4°C in an appropriate buffer at physiological pH. Addition of antimicrobial agents such as sodium azide is advisable. Alternatively the adsorbent may be lyophilized after adding dextran or PEG to prevent the beads from irreversibly collapsing. The beads are washed to remove the added dextran or PEG after reconstitution prior to use.

Many affinity adsorbents are available commercially.

Chromatographic technique. After immobilizing the ligand on to the matrix and blocking the residual active groups on the matrix, the adsorbent is packed into a column and equilibrated with appropriate buffer. Since the capacity of the affinity gels are usually high, short, wide columns containing 1–10 ml gel are used. The chromatographic technique consists of four steps, namely, adsorption of the desired component in the sample, washing off unadsorbed components, elution of the bound component and regeneration of the column.

The sample may be subjected to a pre-concentration or pre-fractionation step of precipitation or ion-exchange chromatography, particularly when large samples are handled, in order to enhance the efficiency of affinity chromatography. The sample containing the desired component is brought into contact with the adsorbent bed by pumping it to the head of the column. Sample volumes are typically less than 5% of the bed volume. The unwanted components of the sample are then washed out with the starting buffer of high ionic strength leaving the column loaded with the desired component.

The adsorbed component in the column is then eluted out by altering the binding equilibrium either specifically or non-specifically. Ligand-counterligand (protein) interactions are mostly based on a combination of electrostatic and hydrophobic interactions and the formation or breaking of hydrogen bonds. Such interacting forces may be weakened non-specifically by using suitable eluents and altering the conditions such as pH, ionic strength or dielectric constant in stepwise or continuous gradient manner to enable the dissociation of the counter-ligand (desired component of the sample) from the immobilized ligand. Denaturants or chaotropic substances such as potassium iodide, potassium thiocyanate, potassium cyanate, urea or guanidine hydrochloride have been used to elute out strongly adsorbed proteins capable of regeneration. In the case of immunosorption, the affinity is quite large and the conventional elution process is not helpful. The polarity of the eluent can be decreased by including ethylene glycol to the extent 20–40% in the buffer. A low concentration of a detergent (e.g., lubrol, NP 40 or octylglycosides) may be used to suppress hydrophobic adsorption as well as aggregation, particularly in the case of hydrophobic proteins such as membrane proteins and in antigen-antibody purification. Affinity elution or biospecific elution may also be practiced by including the free ligand or a competitive binding agent in the eluent. The ligand or the competitive binding agent competes for sites on the counter-ligand (solute component) thereby decreasing the capacity factor k' value of the solute. Such specific eluents are used frequently with group specific adsorbents as the selectivity is greatly increased in the elution step. Thus, for example, lectin (solute) bound to immobilized glucosamine may be specifically eluted out by adding glucose or N-acetyl-D-glucosamine to the eluent. In the so-called *reversed role affinity chromatography* the solute and a competing agent (inhibitor) added to the eluent compete for sites on the immobilized affinity ligand as in the case of immobilized lectin column used to purify glycoproteins. The glycoproteins from the sample mixture can be desorbed specifically from lectin columns by elution with competing carbohydrates such as glucose.

For economic reasons, it is important that affinity adsorbents are regenerated particularly when expensive or rare ligand is used. In practice, the extent to which an adsorbent can be reused depends on the nature of the sample (degrading enzymes may be present) and the stability of the ligand and the matrix towards the conditions used during immobilization, elution and regeneration. In most cases, washing with several columns of starting buffer may be sufficient. Washing alternately with buffers of high and low pH, if necessary including detergents or denaturing agents may also be carried out. The most widely used procedure for regeneration is to wash affinity adsorbent column with several column volumes each of 0.1 M tris/HCl buffer (pH 8.5) containing 0.5 M NaCl followed by 0.1 M sodium acetate buffer (pH 4.5) containing 0.5 M NaCl. The column is then equilibrated with starting buffer prior to reuse.

13.3.4 Applications

Purification and recovery/concentration of biomolecules: Several applications using particularly group specific affinity adsorbents in the purification of biomolecules are well documented. Protein A immobilized on Sepharose has been used for binding different immunoglobulins from different sources. The effect of binding of the proteins to the immobilized ligand has been found to vary with pH and ionic strength. Thus the binding of human IgA was found to decrease with decreasing pH in contrast to the binding of human polyclonal IgG which remained constant. In the case of mouse monoclonal antibodies of different IgG subclasses, the binding to Protein A-Sepharose depends on ionic strength with a high ionic strength of 3 M NaCl required for binding. The different subclasses were eluted out with buffer of decreasing pH.

The use of a low molecular weight ligand, S-adenosyl-L-homocystein immobilized on a spacer arm containing AH-Sepharose 4B as an affinity adsorbent is an example of a specific adsorption and specific elution in the purification of a labile protein catechol O-methyltransferase. The enzyme from rat liver has been purified in a two step process involving ion-exchange chromatography and affinity chromatography. The bound enzyme was eluted out from the affinity column by a buffer containing 0.1 mM S-adenosyl-L-homocysteine, leaving behind other proteins bound to the column.

Affinity scavenging: Selective removal of unwanted contaminants from products may be achieved by this approach. Examples include the selective removal of detergents or endotoxins from protein solutions.

Clinical diagnosis: Affinity methods are advantageous in clinical diagnosis because of their high precision, reliability, quickness and freedom from interferences. For example, immobilized boronic acid on agarose is used as a diagnostic tool to separate and quantify glycosylated haemoglobin in diabetic patients.

Analytical applications: High performance liquid affinity chromatography (HPLAC) involves the use of analytical HPLC column containing affinity gel for analytical purposes as a multianalyte affinity system. For example, human serum albumin (HSA) and IgG have been simultaneously identified in a single sample using a dual column system using Protein A and anti-HSA antibodies as the affinity ligands. Another example involves the analysis of ribonucleosides in urine or deproteinized serum using boronic acid affinity column in series with a conventional reverse phase column.

Immunotherapy and immunoassay: An example is the purification of allergens for use in immunotherapy. High performance immuno affinity chromatography (HPIAC) has been used for correct orientation of the immobilized antibody. The avidin-biotin system was used to prepare immunosorbents on glass beads with directed immobilization of the antibodies. Monoclonal antibodies were biotinylated with biotin hydrazine which couples the biotin to the carbohydrate moieties of the antibodies. The biotinylation ensures correct orientation of the antibody as the carbohydrate part of most antibodies is present in the F_c region. The streptavidin form of avidin was immobilized to activated glass beads and the biotinylated monoclonal antibodies were then attached to the streptavidin coated glass beads. The avidin-biotin immobilized glass beads were packed into a column and used as an affinity column for isolating B 27 human leucocyte antigen component from detergent solubilized human leucocyte membranes. The B 27 antigen retained on the column was eluted by buffer with decreased pH or by the use of chaotropic ions. The biotinylated monoclonal antibodies do not elute out because of the strong affinity to the avidin.

13.4 IMMOBILIZATION OF LIGANDS

The different coupling reagents useful in binding different ligands to a chosen matrix include cyanogen bromide, epichlorohydrin, glutaraldehyde, etc. The chemical activation of the different matrices that can be used for immobilizing ligands vary depending on the nature of the functional groups available on the matrix and the nature of the ligand to be immobilized. The activated matrix in most cases is an electrophile ready to react with amines. Hence buffers containing amino groups (e.g., tris buffer) should *not* be used during activation or coupling. Usually 0.1–0.2 M bicarbonate/carbonate or borate buffer is used.

CNBr activation of polysaccharide matrices and coupling of ligands with amino groups: The method is useful for activation of polysaccharide matrices such as beaded agarose or cross-linked dextran and for coupling ligands capable of multi-point attachment such as proteins and polypeptides. The method is not ideal for immobilizing low molecular weight amino ligands capable of single point attachment as ligand leakage can occur due to hydrolysis.

Polysaccharide matrices may be activated by cyanogen bromide at pH 11–12 to give a variety of products such as reactive cyanate ester and less reactive cyclic imidocarbonate groups in the matrix. In addition, acyclic imidocarbonate, carbamate and carbonate side products are formed. The principal reactive group is the cyanate ester which reacts quickly in mildly alkaline conditions (pH 9–10) with primary amines or through ε-lysine groups of proteins to form isourea derivative. The isourea substituent is positively charged and may introduce an anionic exchange character to the matrix.

$$\text{Matrix}\begin{cases}-OH\\-OH\end{cases} + CNBr \longrightarrow \text{Matrix}\begin{cases}-O-CN\\-OH\end{cases} + \text{Matrix}\begin{cases}-O\\-O\end{cases}\!\!>C=NH$$

Polysaccharide matrix — Cyanate ester — Cyclic imidocarbonate

In the case of agarose a relatively higher concentration of cyanate ester groups are formed while in the case of cross-linked dextran, the imidocarbonate groups are formed at a higher concentration. The activated matrix can be stored for a long time as a suspension or as a lyophilized powder. In the second step, the activated matrix is treated with an aqueous solution of the protein ligand containing a free amino group at a pH of 7–8 to couple the ligand covalently and yield affinity adsorbent.

$$\text{Matrix}\begin{cases}-O-\overset{NH_2^+}{\overset{\|}{C}}-NH-\text{Ligand}\\-OH\end{cases} \qquad \text{Matrix}\begin{cases}-O\\-O\end{cases}\!\!>C=N-\text{Ligand}$$

Isourea derivative — Substituted imidocarbonate

The electrophilicity of CNBr is increased by forming cyanotransfer complex with a base such as triethylamine (TEA) or dimethylamino pyridine (DAP). The cyanotransfer complex formed by DAP is quite stable and is commercially available as the tetrafluoro borate salt, CDAP [1-cyano-(4-dimethylamino)pyridinium tetrafluoro borate]. The use of cyanotransfer agents such as CDAP facilitates the cyanylation of the matrix (particularly beaded agarose) at much lower pH range, but in acetone-water mixture to yield the reactive cyanate ester groups predominantly. It also prevents the hydrolysis of CNBr and wastage of the reagent. In addition, it is easy to handle CDAP unlike CNBr which is highly hazardous.

Epichlorohydrin activation of polysaccharides and coupling of ligands with amino, hydroxyl and thiol groups: Epichlorohydrin activates the polysaccharide matrices at pH 11–12 by introducing three carbon atom propane-2-ol hydrophilic spacer arm containing oxirane groups besides forming crosslinks in the matrix. Epoxy activated matrices react quickly with sulphydryl groups of thiol ligands or primary amine groups of amino ligands and relatively slowly with hydroxyl groups of ligands generating a thioether, a secondary amine or a ether linkage respectively.

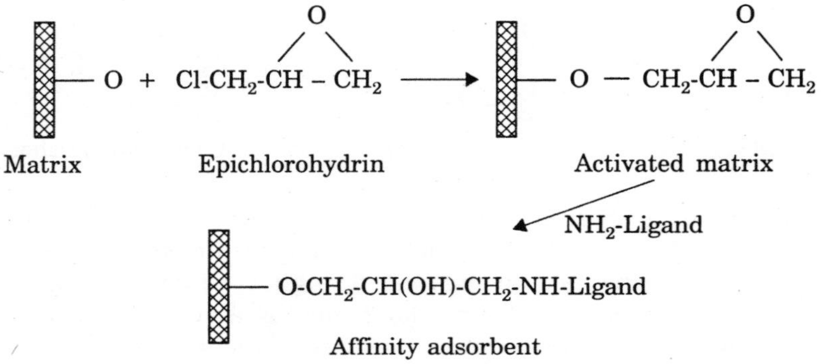

$$\text{Matrix}-O + Cl\text{-}CH_2\text{-}\overset{O}{\widehat{CH - CH_2}} \longrightarrow \text{Matrix}-O - CH_2\text{-}\overset{O}{\widehat{CH - CH_2}}$$

Matrix   Epichlorohydrin   Activated matrix

NH_2-Ligand

$$\text{Matrix}-O\text{-}CH_2\text{-}CH(OH)\text{-}CH_2\text{-}NH\text{-Ligand}$$

Affinity adsorbent

Bisepoxirane activation of polysaccharides and coupling of ligands with amino, hydroxyl and thiol groups: Bisepoxirane (e.g., 1,4-butanediol diglycidyl ether) activation of polysaccharide matrices is similar to that of epichlorohydrin activation but has the advantage of a longer hydrophilic spacer arm containing affinity adsorbent

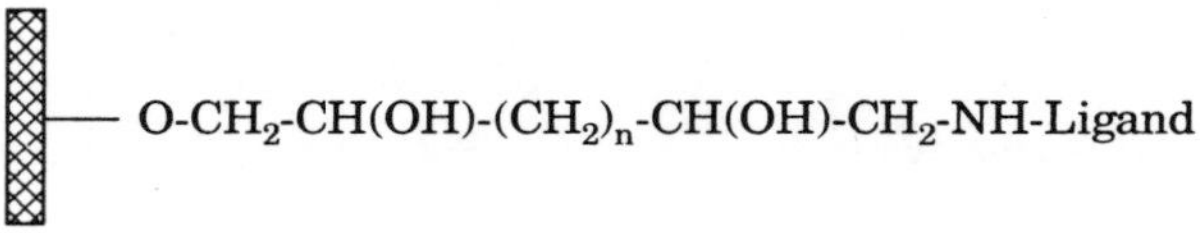

$$\text{Matrix}-O\text{-}CH_2\text{-}CH(OH)\text{-}(CH_2)_n\text{-}CH(OH)\text{-}CH_2\text{-}NH\text{-Ligand}$$

Divinylsulphone activation of polysaccharides and coupling of ligands with amino, hydroxyl and thiol groups: Divinyl sulphone (DVS) ($H_2C{=}CH\text{-}(O)S(O)\text{-}CH{=}CH_2$) activated matrix reacts with amino and thiol groups and slowly with hydroxyl groups of ligands. The activated gels are useful for the synthesis of thiophilic adsorbents and for coupling proteins, particularly immunoglobulins. However, the adsorbents are less stable in alkaline conditions.

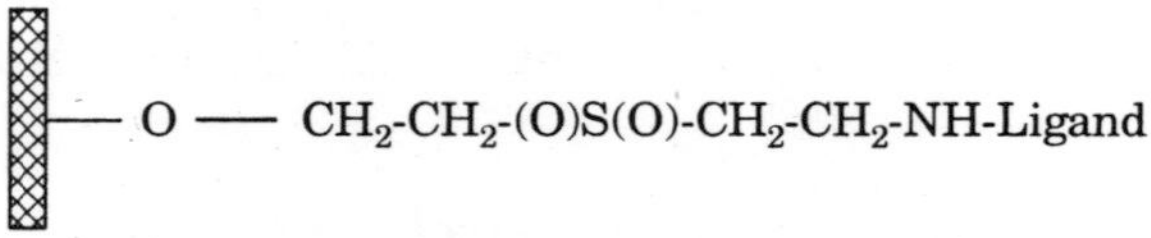

$$\text{Matrix}-O - CH_2\text{-}CH_2\text{-}(O)S(O)\text{-}CH_2\text{-}CH_2\text{-}NH\text{-Ligand}$$

Other methods of activation of polysaccharides include the use of toluene sulphonyl chloride (tosyl chloride) 3,3,3-trifluoroethane-sulphonyl chloride (tresyl chloride) for coupling amino and thiol ligands, NN-disuccinimidyl-carbonate (DSC) and carbonyldiimidazol (CDI) for coupling amino ligands as activating reagents.

Matrices containing carboxyl groups can be activated by carbodiimides for coupling ligands with carboxyl or amino groups.

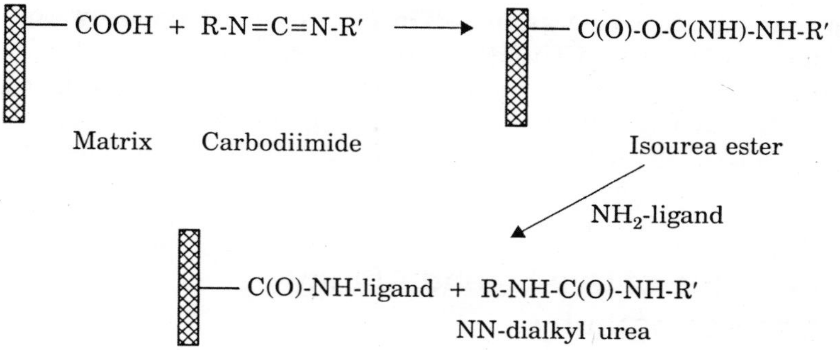

$$\text{Matrix—COOH} + \text{R-N=C=N-R}' \longrightarrow \text{Matrix—C(O)-O-C(NH)-NH-R}'$$

Matrix Carbodiimide Isourea ester

$\downarrow$ NH_2-ligand

Matrix—C(O)-NH-ligand + R-NH-C(O)-NH-R′

NN-dialkyl urea

Polyacrylamide and polyamide matrices can be activated by glutaraldehyde or hydrazine for coupling amino ligands.

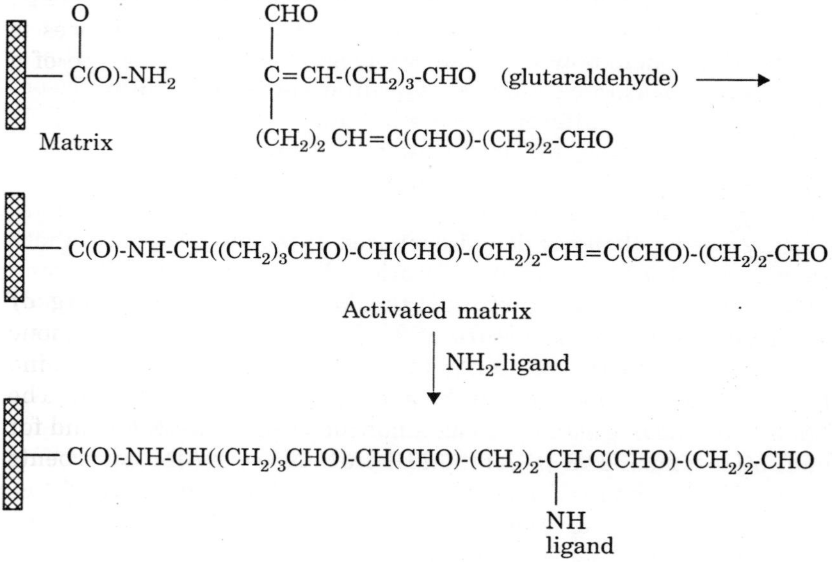

Matrix—$C(O)$-NH_2 (O above C) + CHO—C(=CH-$(CH_2)_3$-CHO)—$(CH_2)_2$ CH=C(CHO)-$(CH_2)_2$-CHO (glutaraldehyde) $\longrightarrow$

Matrix—C(O)-NH-CH((CH_2)$_3$CHO)-CH(CHO)-(CH_2)$_2$-CH=C(CHO)-(CH_2)$_2$-CHO

Activated matrix

$\downarrow$ NH_2-ligand

Matrix—C(O)-NH-CH((CH_2)$_3$CHO)-CH(CHO)-(CH_2)$_2$-CH-C(CHO)-(CH_2)$_2$-CHO, with NH ligand on CH

Silica gel and porous glass matrices are activated by silanizing reagents such as γ-aminopropylsilane or γ-glycidoxypropylsilane followed by coupling ligands with amino, hydroxyl or thiol groups.

13.5 PSEUDOAFFINITY CHROMATOGRAPHY

The affinity of certain nonbiological ligands towards biomolecules has been exploited in the separation and purification of biomolecules. The specificity of such ligands qualifies them as affinity adsorbents. However, in order to distinguish them from natural ligands and their biospecific interactions they may be classified as pseudoaffinity adsorbents. Two well known examples of pseudoaffinity adsorbents include dyes and immobilized metal ions and the chromatographic techniques involving them are named as **dye-ligand affinity chromatography** and **immobilized metal ion affinity chromatography**. Pseudoaffinity chromatography offers an alternative separation methodology using robust and less expensive substitutes for the fragile and more expensive biological ligands in affinity separation/purification of desired biomolecules.

13.5.1 Principle of Dye-Ligand Affinity Chromatography

Textile dyes belonging to a group of polyaromatic sulphonated compounds containing triazine functional group or a vinyl sulphone group, called reactive dyes, are capable of coupling to polyhydroxy matrices. The reactive dyes also interact with a wide variety of proteins to different extents. Because of their polyfunctional groups they mimic to a certain extent the natural nucleotide co-factors. The differences in the affinities of immobilized dyes for different proteins form the basis of the chromatographic separation and purification of proteins.

Dye ligand chromatography came into vogue in 1960s when dyes such as the blue dye known as Cibacron Blue F3G-A were found to exhibit affinity to enzymes particularly dehydrogenases. It is now recognized that most enzymes, which bind a purine nucleotide show affinity for Cibacron Blue F3G-A. The planar ring structure of the dye with negatively charged groups is analogous to the structure of NAD. X-ray diffraction (XRD) studies have shown that the dye binds to liver alcohol dehydrogenase in an NAD site with correspondences of the adenine and ribose rings but not the nicotinamide. The dye behaves as an analogue of ADP-ribose and binds to the nucleotide fold found in ATP, AMP, IMP, NAD, NADP and CTP binding sites. The binding is fairly strong with K values around 10^6. The dye readily couples to the hydroxyl groups of agarose through the triazinyl groups in slightly alkaline conditions. The dye coupled agarose beads serve as excellent affinity adsorbents for a variety of proteins. Other triazinyl dyes such as Procion Red H-E3B (reactive red 120) Procion Turquoise H-A, Procion Green H-E4BD, Procion yellow H-A and vinyl sulphone dyes have also been used. Agarose bound blue and the red dyes have been used for purifying both NAD- and NADP-specific dehydrogenases and other proteins.

13.5.2 Dye-Protein Interactions

Dye-protein interactions are quite complex involving non-specific interactions such as ionic, hydrophobic and charge transfer between the dye and the protein molecules as well as specific interactions. The non-specific hydrophobic interactions involve the planar aromatic rings of the dye molecule and the hydrophobic regions on the protein surface. Electrostatic interactions between sulphonic, carboxyl, amino, chloride and the hydroxyl groups of the dye molecule and charged regions on the protein surface are also involved in dye-protein interactions. In addition to these interactions hydrogen bond formation also occurs. Specific interactions between the dyes and proteins involve divalent metal ions such as Cu, Ni, Co, Mn, Zn and to some extent Mg and sulphonate groups on dye molecules. The sulphonate groups resemble phosphate groups and enzymes containing metal ions exhibit preference for such substrates.

13.5.3 Practice of Dye-Ligand Chromatography

Preparation of dye ligand adsorbents: The effectiveness of the adsorbent depends on the matrix used, the structure of the dye and the degree of substitution. The matrix used is invariably beaded agarose because of its various advantages. The most widely used monochlorotriazinyl dyes are Cibacron Blue F3G-A and Procion Red H-E3B. The hydroxyl groups of agarose react with the reactive functional groups (chlorotriazinyl group of triazine type and vinyl sulphone group of Remazol brand) dyes binding the dyes to the matrix. Multiple dye column kits are available commercially for use in a wide range of separations.

Agarose gel is treated with a 0.2% dye solution at pH 10–11 in sodium carbonate or sodium hydroxide medium containing 2% sodium chloride. Monochlorotriazinyl dyes require 2–3 days for coupling at 20–30°C while the more reactive dichlorotriazinyl dyes require only 2–3 hours. The excess reagent is washed off over a filter funnel and the adsorbent gel is stored in a dilute buffer solution (pH 8–9) with antimicrobial agents such as sodium azide. The degree of substitution of the dye is usually in the range 0.3–5 micromoles/ml. Higher degree of substitution (10–20 micromoles/ml) can be achieved by using an amino-substituted matrix which reacts with dyes. An amino substituted matrix may be prepared by activating agarose with epichlorohydrin and then treating with ammonia. Alternatively, a thiol matrix may be prepared by using sodium sulphide instead of ammonia. The thiol matrix also reacts readily with dyes. However, the S-linked dye is prone to hydrolysis in alkaline conditions.

Chromatographic technique: The steps involved include (i) screening to select suitable dye adsorbent (ii) sample application and washing (iii) elution of the bound protein and (iv) regeneration of the dye adsorbent.

The choice of dye adsorbent for a given separation or purification is decided on the basis of a preliminary screening using small scale columns or batch adsorption studies. Screening is carried out in suitable buffers such as *N*-morpholinoethane sulphonic acid (Mes) at pH 6–7 and *N*-morpholinopropane sulphonic acid (Mops), *N*-tris(hydroxymethyl) methyl-2-aminoethane sulphonic acid (Tes) or Tris buffers at higher pH values. A differential mode of chromatography has been suggested for efficient separation and purification of the desired protein. The differential mode of operation involves the use of two separate dye adsorbent columns with opposite differential selectivities in tandem. In the first column the dye adsorbent binds most of the proteins in the sample mixture except the desired protein. The effluent from this column is fed into the second column with a dye adsorbent exhibiting opposite selectivity in that the desired protein alone is bound but not other proteins. This differential mode of chromatography gives rise to a high degree of separation and purification of the desired protein in the adsorption stage. If the bound protein-dye adsorbent system is amenable for affinity elution of the protein, a single step purification of the desried protein becomes possible.

Elution of the dye adsorbent bound protein may be carried out non-specifically or by affinity elution. The dye-protein interactions may be weakened and the protein may be eluted out non-specifically by changes in buffer. Elution may be carried out by increasing the pH including a pH gradient or by increasing hydrophobic interactions between the mobile phase and the bound protein by increasing the salt content or by including solvents such as ethanol, iso-propanol or ethylene glycol or non-ionic detergents in the buffer. Chaotropic salts such as LiBr and KCNS may also be used as in the case of purification of human serum albumin.

The dye adsorbent columns may be cleaned of any bound proteins by washing successively with alkali or 6 M urea in 0.5 sodium hydroxide solution followed by several column volumes of dilute sodium chloride solution and finally with the equilibrating buffer for the next use. The dye absorbent may be stored for a long time in alkaline phosphate buffer containing 10 mM sodium azide. Many dyes leak from agarose and before use the adsorbent may be washed with dilute alkali to remove the leaked dye.

13.6 IMMOBILIZED METAL ION AFFINITY CHROMATOGRAPHY (IMAC)

13.6.1 Principle

Many proteins form multidentate or chelate complexes with transition metal ions such as Fe^{2+}, Ni^{2+}, Cu^{2+} and Zn^{2+} through electron donor groups involving O, N or S atoms. The stability of the complexes varies widely with different proteins and the difference in the affinities of proteins for immobilized metal ions forms the basis of immobilized metal ion affinity chromatography.

The chromatographic matrix consists of a metal chelating ligand such as the tridentate iminodiacetate (IDA), the tetradentate nitrilotriacetate (NTA) or the pentadentate tris(carboxylmethyl)ethylenediamine (TED) immobilized on the support gel by covalent coupling. The immobilized chelating ligand has free coordination sites capable of binding or immobilizing metal ions (e.g., divalent Cu, Ni, Zn, Co, etc.) to form the immobilized metal ion affinity adsorbent. The immobilized metal ion still has unsaturated coordination sites capable of interacting with solvent molecules of the mobile phase or the any sample proteins that may be present in the mobile phase. In the absence of sample proteins the immobilized metal ion is coordinatively unsaturated with solvent molecules loosely binding to the coordination sites. A mixture of proteins bind to the immobilized metal ion to different extents depending on factors such as the presence of different amino acid residues exposed on their surface, steric hindrance, pH of the medium, ionic strength etc. A schematic diagram of the binding of a protein on an immobilized metal ion affinity adsorbed containing IDA is shown in Figure 13.3.

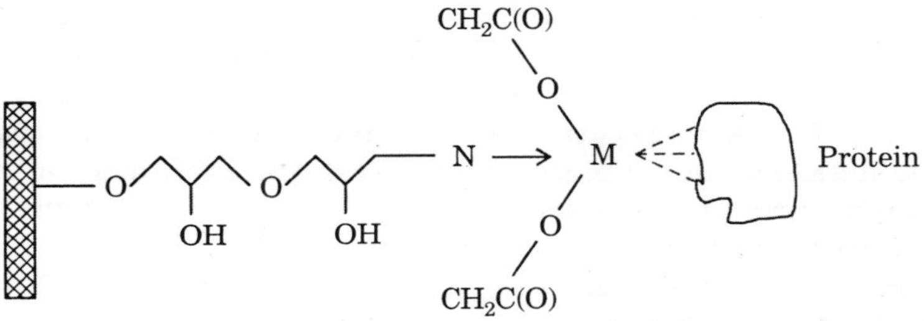

Figure 13.3 Coordinatively unsaturated immobilized metal ion binding to surface amino acid residues of a protein molecule.

The effectiveness of binding of the protein to the immobilized metal ion depends on the exposure of certain amino acid residues such as histidine, cysteine and tryptophan on the surface of the protein. These amino acid residues with free functional groups can coordinate to the immobilized metal ion. The strength of this interaction depends on the number of such coordinative linkages between the metal ion and the protein. For example, it has been shown that a single histidine moiety in a small protein is sufficient to bind to an immobilized metal ion with K value in the range of 10^6. The nature of the metal ion is also important in deciding the strength of interaction. For example, ferric iron has been shown to be selective for phosphorylated proteins and is useful for separating the phosphorylated and unphosphorylated forms of the same protein. In practice other factors such as steric considerations between the potential coordination sites on the surface of the protein, i.e., the secondary and tertiary structure of the protein and external factors such as pH and ionic strength of the medium also affect the strength of binding. Hence it is possible to separate proteins of close similarities with respect to charge, molecular size and amino acid composition based on the differences in their secondary and tertiary structures. An inherent advantage of the technique is its ability to concentrate the desired protein component in the sample.

13.6.2 Practice of IMAC

The IMAC gel is conveniently prepared by activating the beaded agarose or crosslinked dextran matrices with epichlorohydrin and coupling with the chosen chelating ligand in alkaline medium. IMAC adsorbents without the metal are available commercially. The chosen metal ion is immobilized by simply equilibrating the metal ion solution with the adsorbent gel and washing off the excess of unadsorbed metal ion with water followed by washing with 0.5 M NaCl solution to remove any physically occluded metal ion and finally with a buffer chosen for elution at a later stage. Any weakly bound metal ion may be removed by washing the gel with a solution of a weak complexing ligand such as imidazole or glycine prior to washing with buffer. The immobilized metal ion may be stripped off the gel and completely washed out by equilibrating and washing with a strong solution of EDTA. The gel is now ready for loading a different metal ion. Most widely used metals include divalent zinc, copper and nickel.

The IMAC gel is packed into a column and equilibrated with buffer. The sample solution in the starting buffer is preferably dialysed to remove any interfering solutes before its application to the column. Different elution techniques such as isocratic, continuous or stepwise gradient of pH, or competitive elution using a buffer containing a chelating ligand such as glycine, histamine, histidine or imidazol may be used. In all the elution modes the ionic strength is maintained at relatively high level to minimize non-specific adsorption of the solutes. The column may be regenerated after elution of the desired component by washing with a buffer if the same metal ion is to be retained or with a strong solution of EDTA if the metal ion-free IMAC gel is required for subsequent use or storage.

A few applications are cited here to show the importance of this technique. Human plasma proteins α_2-macroglobulin (α_2-M) and α_1-antitrypsin (α_1-AT) have been purified by IMAC using a dialyzed sample obtained by ammonium sulphate fractional precipitation. The protein α_2-M was retained on a Zn-IDA column at pH 6 while the contaminating proteins were washed off the column. The bound protein was eluted at pH 5. The protein α_1-AT was purified from human plasma by retaining on Zn-IDA column at pH 8 while the major contaminating plasma protein albumin was not retained on the column. The bound α_1-AT was eluted out at pH 6.5 with a 20 fold purification. IMAC has been useful in the purification of several types of interferons. Human fibroblast interferon was purified using Zn-IDA column.

13.7 COVALENT CHROMATOGRAPHY

Covalent chromatography, as the name implies, involves the formation and breaking of covalent bonds between the solute component of a mixture and the chromatographic stationary phase under specific and mild conditions. Among the various functional groups in proteins amenable for chemical reactions amino and carboxyl functional groups form relatively strong bonds and hence are mostly used for immobilization of the protein on a matrix. Any attempt to release the immobilized protein is more likely to destroy the protein. However, the thiol (RSH) groups form covalent bonds which can be split under mild conditions without destroying the protein. Covalent chromatography thus uses the feature of reversible covalent bond formation for the isolation and purification of thiol containing biomolecules. Covalent chromatography has been extended to substances containing methionine and tryptophan groups though it is practiced less frequently.

13.7.1 Chemistry of Thiol Group

Thiol groups of proteins participate in a variety of chemical reactions such as dissociation, oxidation, complex formation with metals and alkylation. Thiols undergo dissociation to give the highly nucleophilic thiolate ion (RS^-) in neutral and mildly alkaline conditions. The thiol groups undergo oxidation to form disulphide (RSSR) which are relatively less reactive but important for the stabilization of the tertiary and quaternary structure of the proteins. The disulphide group undergoes further oxidation at higher pH values and in the presence of oxidizing agents to give sulphenic (RSOH), sulphinic (RSO_2H) and sulphonic (RSO_3H) acids. The ability of the thiol groups to form complexes with metal ions such as copper, zinc and mercury has been used in covalent chromatography by using stationary phase gel immobilized organomercury compound. Alkylation of thiol to give a thioether may be used for blocking the reactive thiol group.

The reversible thiol-disulphide exchange reaction is actually S-alkylation and is of special significance in covalent chromatography. The two step nucleophilic displacement in which a mixed disulphide intermediate is formed, may be considered as a redox reaction involving the oxidation of thiol to disulphide and the reduction of the disulphide to thiol.

$$R_1SSR_1 + RSH \longrightarrow R_1SH + R_1SSR$$

$$R_1SSR + RSH \longrightarrow R_1SH + RSSR$$

or

$$R_1SSR_1 + 2RSH \longrightarrow 2R_1SH + RSSR$$

The exchange reaction is faster at higher pH values and since the equilibrium constant for this redox reaction is nearly unity, a large excess of thiol is required to reduce the disulphide and similarly a large excess of disulphide is required to oxidize the thiol.

13.7.2 Biological Importance of Thiol and Disulphide Groups

A number of reactions within the biological cells are critically dependent on the free thiol groups. The intracellular environment is kept in a reduced state by cysteine containing tripeptide glutathione [GSH = γ-glutamyl-cysteinyl-glycine, $-OOC\text{-}CH(NH_3^+)CH_2CH_2CONHCH(CH_2SH)CONHCH_2COO^-$]. GSH is present in millimolar concentration and accounts for nearly 90% of all the thiols in the cell. It serves as a scavenger of free radicals and as an oxidant with molecules capable of reacting with thiol groups. The high concentration of GSH is maintained by biosynthesis through glutatione synthetase and also by the reduction of the disulphide group in the oxidized glutathione (GSSG). The reduction is catalyzed by the NADPH dependent glutathione reductase. In an erythrocyte the ratio of GSH/GSSG is about 500. The thiol-disulphide exchange reaction is important in the biosynthesis of proteins, aggregation and polymerization of some proteins and possibly the regulation of activities of some intracellular enzymes, effector-receptor interactions and membrane transport.

Intracellular proteins have mostly free thiol groups and contain very few disulphide groups. The free thiol groups of enzymes are known to participate in catalytic reactions involving oxido-reductases, transferases and intracellular proteases (cathepsins) and alkylation of the thiol groups inactivates the enzymes. In alcohol dehydrogenase, the thiol group is known to serve as a ligating site to complex metal ions. In contrast, extracellular proteins with disulphide groups are more commonly found and thiol groups occur only when they are required for a special purpose. Examples of extracellular enzymes include serine proteases trypsin and chymotrypsin and blood coagulation factors all of which have disulphide bridges but not thiol groups. However, plant proteases such as ficin, bromelain and papain are extracellular thiol dependent enzymes. The thiol groups in these enzymes are protected against oxidation within the active site. Albumin, the mammalian plasma protein sometimes carries a free thiol group and traps thiol or disulphide containing compounds in the serum.

13.7.3 Principle of Covalent Chromatography

The principle is based on facilitating thiol-disulphide exchange reaction between gel immobilized thiol groups and the thiol containing proteins in the mobile phase is the first step to covalently bind thiol proteins specifically to the gel through the formation of disulphide bridge. After washing off the unbound (non-thiol) proteins, the bound thiol protein is subjected to thiol-disulphide exchange reaction once again by using a mobile phase containing reducing low molecular weight thiol compounds. Thus, covalent disulphide bridges are formed during the binding of the thiol proteins to the gel and broken during elution step to release the thiol proteins. The two step process of immobilization of the thiol protein and release of bound protein are schematically shown below.

[gel]—S-S-R + HS-Protein ⟶ [gel]—S-S-P + RSH

Activated thiol gel   Thiol protein   Bound thiol protein

[gel]—S-S-P + HS-R (excess) ⟶ [gel]—SH + HS-P + R-S-S-R

13.7.4 Practice of Covalent Chromatography

The steps involved include (i) preparation of the stationary phase gel (ii) sample preparation (iii) binding of the thiol proteins (iv) washing off the unbound proteins (v) reductive elution of the bound thiol proteins from the column (vi) recovery of thiol proteins and (vii) reactivation of the thiol gel.

Preparation of stationary phase gels. The matrices that can be used for preparation of gels for covalent chromatography include agarose, cross-linked dextran, cellulose, polyacrylamide, porous glass and silica. The matrix chosen for gel preparation must have the same characteristics as applicable for other chromatographic techniques. Beaded agarose has been widely used as the matrix because of its advantageous features. In the first step thiol groups are introduced into agarose gels and the gels are then activated by reacting with a reactive disulphide such as 2,2′-dipyridyldisulphide (PDS) making them amenable for use as the stationary phase in covalent chromatography. Thiol groups may be introduced into agarose gels by several methods. One of the methods uses coupling of glutathione to cyanogen bromide activated agarose gels.

Another method for the introduction of the thiol groups in agarose gels involves the reaction of the epichlorohydrin or bis-epoxide activated agarose with sodium thiosulphate to give thiopropyl gel which, in turn can be activated by reacting with PDS as shown below.

[gel]—OCN + H_2N-glutathione-SH ⟶ [gel]—OC(=NH_2)–NH-glutathione-SH

Activated matrix   Thiol containing gel

— $OC(=NH_2)$–NH-glutathione-SH + N(pyridyl)-S-S-(pyridyl)N →

— $OC(=NH_2)$–NH-glutathione-S-S-(pyridyl)N
Activated gel

The degree of substitution with respect to the content of active 2-pyridyldisulphide (PyS-S) groups is quite low in the case of glutathione coupled gel while it is very high (almost 100 times more) in the thiopropyl coupled gel. The degree of substitution is determined by reacting the activated gel with a low molecular weight thiol and measuring the thiopyridone by spectrophotometry at 343 nm. The dynamic capacity under chromatographic conditions is much smaller compared to the degree of substitution.

Other reactive disulphides which may be used for activating thiol gels include (5-carboxy-2-pyridyl)disulphide and 5,5′-dithiobis-(2-nitrobenzoate). Thiol gels such as cysteamine agarose (matrix-$OC(=NH^{2+})NH$-CH_2CH_2SH), cysteineagarose (matrix-$OC(=NH^{2+})NHCH(COOH)$-CH^2 SH), *N*-acetylhomocysteineagarose (matrix-spacer-$CH(NHCOCH_3)CH_2$ CH_2SH) and glutathioneagarose (matrix-glutatione-SH) and PDS activated thio gels containing glutathione or thipropyl groups are commercially available.

The glutathione activated gel is negatively charged because of the presence of the carboxylate anions while epicholohydrin activated gels are neutral. Both the gels can be stored as suspensions at 4°C and pH 6–7 for several months or in lyophilized form. Sodium azide should *not* be used as it is a good nucleophile and deactivates the gel.

Sample preparation. The sample is dialysed or passed through gel filtration media to remove any low molecular weight thiol compounds such as glutatione that may be present. The thiol content of the sample is preferably determined by reacting a small portion of it with PDS and determining the 2-thiopyridone released spectrophotometrically at 343 nm. Buffers in the pH range 3–8 such as formate, acetate, phosphate and tris containing 0.05 M EDTA to chelate any metal ion may be used with or without denaturing agents such as 8M urea or 6M guanidine hydrochloride.

Binding of proteins. The activated gel is packed into preferably a narrow bore column and the sample is allowed to remain in contact with the gel for about an hour to enable binding of the thiol proteins in the sample to the gel. The effluent from the column may be spectrophotometrically analysed at 343 nm to determine the amount of reacted thiol groups in the matrix.

Washing the column. The column is then washed with 2–3 column volumes of a high ionic strength buffer containing 0.1–0.3 M NaCl and if necessary with detergents such as Triton and Tween to remove unbound and non-specifically adsorbed proteins in the sample.

Reductive elution. The gel bound thiol protein is eluted out by reductive elution using a buffer at pH 8 containing an excess of a low molecular weight thiol compound such as dithiothreitol (DTT) and 2-mercaptoethanol. Cysteine may also be used for reductive elution; but its oxidized form, cystine is less soluble and may form a precipitate. The elution process may be followed by measuring absorbance at 343 nm. The reductive elution may be carried out in two steps to avoid contamination of the sample protein with thiopyridone from unused pyridyldisulphide groups on the gel. In the first step equimolar amount of the reducing agent in acidic or alkaline pH is used to remove the contaminants followed by washing the column and then eluting out the bound protein with excess of reducing agent in the second step. Sequential elution of bound proteins may also be carried out using thiol compounds of differing reducing power.

Recovery of thiol proteins. The released thiol protein from the column obtained by a single step reductive elution contains thiopyridone and excess of reducing agent which are removed by gel filtration preferably using a volatile bufffer at low pH to facilitate the lyophilization of the recovered thiol protein.

Reactivation of gel. The chromatographic gel after its use is in the reduced thiol form and has to be activated before its reuse. The gel is incubated for about 45 minutes with a 5 mM solution of DTT to reduce the aliphatic disulphides formed on the gel during elution. The excess of DTT is then washed off with a buffer. The gel is then reactivated by incubating for 45 minutes with a saturated (1.5 mM) solution of PDS in 0.1 M phosphate buffer (pH 8). High capacity gels may be reactivated using a 20 mM solution of PDS in a buffer containing 20–30% ethanol to dissolve PDS. The gel is washed to remove excess of PDS and can be stored at 4°C.

Applications. Thiol containing proteins are easily separated from non-thiol proteins and purified by this technique. By using sequential elution with different thiol reducing agents a high degree specificity is easily achieved as in the following case.

Thiol rich urease (500 kDa) from jack bean meal was purified by covalent chromatography. The crude extract of jack bean meal in 0.05 M tris-HCl buffer adjusted to pH 7.2 was chromatographed on a high capacity thiopropylagarose gel column equilibrated with 0.05 M tris-HCl buffer, pH 7.2 containing 0.1 M KCl and 1 mM EDTA. The unbound proteins were washed off the column with tris-HCl buffer. The gel bound urease was eluted out using 20 mM DTT in 0.05 M tris-HCl buffer of pH 8 containing 0.1 M KCl and 1 mM EDTA. A 280 fold increase in the urease activity compared to that of the starting extract was obtained after the removal of the low and high molecular weight substances by gel filtration of the chromatographed material.

Papain (23.5 kDa) contains only one thiol groups which is essential for its activity. The crude papain extract obtained from papaya latex in tris or acetate buffer was purified on a low capacity thiol agarose gel. The sample dissolved in 0.1 M Tris-HCl buffer at pH 8 containing 0.3 M NaCl and 1 mM EDTA was applied to activated thiol agarose column washed and equilibrated with the application buffer. The bound papain was eluted out using 50 mM L-cysteine in the same buffer. Further purification of the enzyme was carried out by ammonium sulphate (30% saturation) precipitation followed by gel filtration of the redissolved precipitate in pH 8 buffer containing 0.1 M KCl and 1 mM EDTA on Sephadex G-25 column.

Covalent chromatography in conjunction with sequential reductive elution with different thiol compounds of increasing reducing power is the only method by which the two enzymes, protein disulphide isomerase (PDI) and glutathione-insulin transhydrogenase (GIT) have been separated and purified from other thiol proteins in beef liver extract. The protein sample from the extract was partially purified by ion-exchange chromatography. The sample was then pretreated with 0.1 mM DTT in TKM/EDTA/NaCl buffer (50 mM Tris-HCl pH 7.5 buffer containing 25 mM KCl, 5 mM $MgCl_2$/1.25 mM EDTA/0.1 M NaCl) at 30°C for 30 minutes to unmask any buried mixed disulphide groups in the proteins. The pretreated sample was centrifuged and subjected to Sephadex G-25 gel filtration to remove DTT. The reduced protein sample was adsorbed on activated high capacity thiopropyl agarose gel by incubating for about 16 hours. The gel containing the bound proteins was then packed into a column, cooled to 4°C, washed with TKM/EDTA/NaCl buffer to remove non-specifically bound and unbound proteins in the sample. The bound thiol proteins were then eluted sequentially using 20 mM L-cysteine, 50 mM glutathione and 20 mM DTT each in the same buffer. The bulk of the enzyme PDI (70–98%) was eluted out by L-cysteine buffer while GIT eluted out with DTT buffer (89–100% yield).

In other applications of covalent chromatography, free thiol group containing family of glycoproteins called Band 3 proteins from human erythrocyte plasma which are involved in the transport of anions, cations and possibly glucose have been isolated. Isolation of thiol peptides from protein digests and characterization of subunits of thiol proteins are other examples. For instance, urease from jack bean has been shown to be a hexameric enzyme, with each subunit being active in isolated immobilized form contributing to 1/6 the activity of native urease.

Questions

1. Distinguish between specific and non-specific interactions in chromatography.
2. Discuss the principle of bioaffinity chromatography.
3. What are the advantages of bioaffinity chromatographic technique?
4. Explain the concept of spacer arm and its use.
5. Write a note on the different ligand immobilization techniques.
6. What are the characteristics of chromatographic support matrices for use in bioaffinity chromatography?
7. What is pseudoaffinity chromatography?
8. Write a note on dye-protein interactions.
9. How is dye-ligand chromatography useful in the purification of proteins?
10. Discuss the principle and practice of immobilized metal ion affinity chromatography.
11. Write a note on the reactions of thiol groups.
12. How are stationary phase matrix gels for covalent chromatography prepared?
13. Explain the principle of covalent chromatography. What are the applications of the technique?

CHAPTER

14

Electrokinetic Methods of Separation

14.1 THE VARIOUS METHODS

Electrokinetic methods of separation include electrophoresis and related techniques, isoelectric focusing and isotachophoresis. Electrophoresis is performed by applying a constant electric field at a constant pH to separate charged species on the basis of their mobility. Capillary electrophoresis is a related technique which has gained importance as an analytical tool capable of high resolution. Isoelectric focusing uses a constant electric field in conjunction with a pH gradient so that charged species migrate till they become electrically neutral and get concentrated or focused at their respective isoelectric points (pI values). Isotachophoresis separates charged species in an electric field of different strengths in conjunction with a pH gradient on the basis of their mobility per unit electric field.

14.2 ELECTROPHORESIS

Electrophoresis can be considered as an incomplete form of electrolysis. The applied electric field is removed before the sample molecules reach the opposite electrodes and the molecules get separated based on the differences in their electrophoretic mobility. The separation of charged molecules or species is brought about by their differential migration under the influence of an applied electric field. Differential migration of solutes and hence their separation is due to the differences in solute charges and diffusion coefficients. Many of the biomolecules such as amino acids, peptides, proteins, nucleotides and nucleic acids possess ionizable groups and exist in aqueous media as charged species depending on the pH of the medium. Even typically non-polar substances, such as carbohydrates can be made weakly charged by derivatization as borates or phosphates.

Molecules may have similar charge but still may be differentiated on the basis of their charge/mass ratios due to their inherent differences in molecular weights. The combination of these differences results in differential migration of the charged species when they are subjected to an electric field and forms the basis of electrophoretic separation.

14.2.1 Basic Concepts

In an applied electric field, an attractive force is exerted on the charged molecules or particles making the molecules to accelerate towards oppositely charged electrodes. If the net charge on the molecule is Q and the electric field strength is E, then the force responsible for the migration of the molecule is equal to the product QE. The acceleration caused by this electrophoretic attraction is opposed by frictional force F which is given as $F = fv$, where f is the friction coefficient and v is the migration velocity of the molecule. Assuming Stokes' law for spherical molecules of radius, r, at very low Reynolds numbers, moving through a liquid medium of viscosity η, the friction coefficient f is given by:

$$f = 6\pi\eta r \tag{14.1}$$

At steady state,

$$QE = fv \quad \text{or} \quad QE = 6\pi\eta r v \tag{14.2}$$

The electrophoretic mobility of the charged species v' is defined as the migration per unit field strength.

$$v' = \frac{v}{E} = \frac{Q}{6\pi\eta r} \tag{14.3}$$

Since $Q = ze$, where z is the valence and e is the electron charge (1.602×10^{-19} coulombs), the above equation becomes,

$$v' = \frac{ze}{6\pi\eta r} \tag{14.4}$$

where E is given in volts/m or volts/cm, viscosity in Pascal second (or Ns/m^2) and electrophoretic mobility in m^2/volt sec or cm^2/volt sec.

The mobility of the molecule thus depends on the viscosity of the medium, the size and shape (Stokes' radius) and the charge on the molecule. In the case of macromolecules the electrophoretic mobility depends mainly on the ionizable groups present on the surface of the molecule and the sign and magnitude of the charge carried by the ionizing group. The sign and magnitude of the charge vary depending on the ionic strength and pH of the medium. Separation of molecules can therefore be effected by selecting an appropriate medium.

Equation (14.4) actually gives the electrophoretic mobility of an isolated species which is its intrinsic property under defined conditions of temperature, buffer and pH and composition of the medium. However, for a large number of colloidal charged particles in an aqueous medium, the real mobility of the species depends on three other retarding forces. These are: (i) electrophoretic retardation (ii) relaxation effect and (iii) electro-osmotic flow of the solvent. Electrophoretic retardation is due to the presence of a layer of liquid, one or two molecules thick, bound to the charged species retarding the molecule as it moves. The electrical potential at the surface of the charged species is slightly lowered to a value called the zeta potential (ζ) which controls the actual mobility of the charged species. Usually the zeta potential is determined by measuring the electrophoretic mobility. The relaxation effect is due to an ion atmosphere formed by oppositely charged species surrounding each charged species. In an applied electric field, the charged species and the surrounding ion atmosphere move in opposite directions resulting in a distorted ion atmosphere and relaxation effect. Electroosmotic or electroendosmotic flow refers to the flow of bulk liquid solvent under the influence of electric field. Each charged species is associated with a number of coordinated water molecules, which also migrate along with the ion. However, immobilized negative charge groups on the walls of the electrophoresis channel and in any stabilizing gel cannot migrate but the counter cations move towards the cathode causing the flow of water in the same direction.

14.2.2 Electrophoretic Techniques

Different types of electrophoresis have come into vogue depending on whether the electrophoresis is conducted in the presence of a supporting medium or not. Electrophoresis in the absence of any supporting medium is called *free solution electrophoresis* whereas if the technique is carried out using a support medium it is known as *zone electrophoresis*.

Free boundary or moving boundary electrophoresis technique carried out in free solution was first introduced by the Swedish biochemist Tiselius. In free solution electrophoresis, the sample is dissolved or suspended in a buffer for electrophoresis to take place. The current is maintained by electrolysis occurring at the two electrodes dipping in the buffer reservoir. The bulk of the current is carried by the buffer ions in solution with only a small component of current being conducted by the sample molecules. The buffer also maintains a constant state of ionization since changes in pH would alter the charge on molecules being separated, particularly when the molecule is zwitter ionic. The technique has several disadvantages in that the separation of proteins is incomplete due to mixing of component bands. Factors such as diffusion of the components, protein-protein interaction and electrostatic interaction cause mixing and overlapping of the sample components and affect the resolution. Electrostatic interaction can be minimized by increasing the ionic strength of the buffer medium but this causes higher current flow during electrophoresis and consequently more heat is generated. The nature of protein-protein interaction depends on the hydrophilic or hydrophobic nature of the proteins, it being predominantly electrostatic for hydrophilic proteins whereas in the case of globular proteins it is more of hydrophobic nature.

In the Tiselius method, the sample protein solution is dialysed against a buffer solution to get sharp boundaries between the protein solution and the buffer in each arm of a U shaped tube. Electrophoresis is then carried out and the movement of the proteins at the two boundaries is monitored by measuring the refractive index of the solution. The method was useful in studying the effect of the physical properties of the solution on the mobility of proteins. However, complete separation of proteins could not be achieved due to difficulty in maintaining stable boundaries. Mixing of the boundaries occurs due to diffusion as well as convection.

Zone electrophoresis involves the use of inert and relatively homogeneous supporting medium impregnated with buffer solution and has distinct advantages over free solution electrophoresis in that mixing due to convection or diffusion of components and also the heat generated during electrophoresis are negligible. Electrophoresis of a sample on a supporting medium causes its components to migrate as distinct zones, which can be subsequently detected by suitable analytical techniques. The technique finds extensive use in analytical as well as in preparative work. The various supporting media used in zone electrophoresis include a sheet of absorbent paper (paper electrophoresis), cellulose acetate film or a thin layer of silica or alumina or a gel of starch, agar or polyacrylamide (gel electrophoresis). All the supporting media have a capillary structure with good anticonvectional properties. Of these gel electrophoresis has found wide use in analytical as well as in preparative work.

Gel electrophoresis is one of the widely used forms of zone electrophoresis using a gel of starch or polyacrylamide to separate species on the basis of both charge and size. It is possible to separate proteins and oligonucleotides of even similar electrophoretic mobility but differing in their size (or molecular weight). The gel is a three-dimensional network of filaments forming pores of differing sizes and the gel can be designed to give pore size suitable for a particular set of biomolecules being separated. Because of the small pore size of the gel compared to the pore size in other supporting media such as paper or cellulose acetate, a molecular sieving mechanism aids separation of molecules on the basis of their size with large molecules being retarded. The effective viscosity of the gel medium within the molecular sieving range depends on the molecular size of the sample components and the electrophoretic mobility of the components depends on the molecular weight (linear with log M) for similarly shaped molecules. The gel minimizes convectional and diffusional movement of the solute components retaining the sharpness of the band as it moves through the gel resulting in complete resolution of the components.

Polyacrylamide gel electrophoresis was originally developed as 'disc gel electrophoresis' as the sample proteins were separated into discs of protein zones on tubes of gel. The disc gel electrophoresis has been largely replaced by 'thin-slab gel electrophoresis' using polyacrylamide because the thin gel generates less heat during electrophoresis and identification of protein bands with dyes is easier as the dyes diffuse rapidly into the thin gel. Several commercial thin-slab electrophoretic units for casting the gel and conducting the electrophoresis are available. The most commonly used dye is Coomassie Blue. After electrophoresis the gel is soaked in a solution of the dye in dilute acetic acid-ethanol solvent for staining. Excess dye is removed by soaking the gel in the dye-free solvent to give coloured protein-dye bands in a clear background.

The buffer or the suspension medium needs to be carefully selected since some buffer ions react with the compounds under investigation. For example, borate forms complexes with sugars. The bulk of the current flowing through the separation channel is carried by the buffer ions. Ions such as Na^+, K^+, Mg^{2+}, F^-, Cl^-, Br^-, SO_4^{2-} and HPO_4^{2-} are highly mobile and carry much current leading to heating and hence may be avoided unless specifically required. Bulky organic ions have much lower mobility. The composition of the buffer and its ionic strength is chosen on the basis of a compromise between low conductivity (hence low heating) which allows application of a higher voltage but may result in protein-protein interactions on the one hand and on the other hand, a buffer system with higher conductivity which minimizes protein-protein interaction but applied voltage must be minimum to avoid heating. The pH chosen depends on the particular mixture of proteins under investigation; but generally maximum separation of components is obtained at the isoelectric point of one of the components. The selected pH should not cause any chemical changes or denaturation of the molecules under study.

In practice, the buffer systems used for electrophoresis may be continuous or discontinuous systems. In the continuous buffer system, the buffer composition is the same throughout with the electrode chambers having a higher concentration (~0.2 M) to reduce voltage drop and the separation channel having a lower concentration of buffer (~0.05 M). Normally buffer concentrations of total ionic strength in the range 0.05–0.15 M are used. The most widely used continuous buffer system is Tris-borate at pH 8–8.5 without or with a low concentration of EDTA. Other examples of buffer systems include Tris-Tricine (pH 7.5–8), imidazole-Mops (*N*-morpholinopropane-sulphonicacid) (pH 7–7.5). Discontinuous buffer systems are used to sharpen the protein bands, particularly in analytical work. The principle is based on Kohlrausch discontinuity being established by a combination of starting buffer anions of higher mobility (e.g. chloride, citrate, phosphate or EDTAate) and following anions of lower mobility (e.g. borate or glycinate at pH 8–9). The discontinuity junctions is self-sharpening and hence can be observed from refractive index changes. The sharpening of the protein bands occur due to (i) the low mobility immediately behind the junction which facilitates the rapid migration of trailing proteins due to a high local electric field and (ii) a higher pH immediately behind the junction which also results in the sharpening of the protein band.

14.2.3 Practice of Analytical Gel Electrophoresis

Three different methods are commonly used in analytical gel electrophoresis. These include (i) simple or native gel electrophoresis (ii) SDS-PAGE and (iii) urea gel electrophoresis.

Native gel electrophoresis. The sample proteins are dissolved in a buffer at a pH where the proteins remain stable and in their native conformation. The pH range is usually 8–9 where most of the proteins carry negative charges and move towards the anode placed at the bottom of the vertical polyacrylamide gel. Any basic protein present in the sample remains dissolved in the cathodic buffer at the top of the vertical gel. A buffer with lower pH range may be chosen and the cathode placed at the bottom of the vertical gel if the sample proteins are basic. Thus, this form of electrophoresis allows the movement of the proteins only in one direction. The gel is cast as a thin vertical slab between two parallel glass plates by polymerizing acrylamide solution (usually 7–10% w/v in buffer) and the cross-linker N,N′-methylene bisacrylamide with ammonium persulphate (~2 mM) together with TEMED (~0.1% v/v N,N,N′, N′-tetramethylethylenediamine) a free radical scavenger. The pore size of the gel can be tailored to suit the molecular weights of the sample proteins by altering the concentration of either the monomer acrylamide or the cross-linker. Increasing the concentration of either of the two decreases the pore size and vice versa. For small proteins of about 10 kDa, a gel containing 20% acrylamide and 1% bisacrylamide may be used while for large proteins up to 1000 kDa, a gel of 4% acrylamide and 0.1% of bisacrylamide may be used. Once the gel is formed, in about 30 minutes, the sample proteins dissolved in the chosen buffer are applied in the wells set in the gel using a 'comb'. A schematic diagram of the electrophoretic unit is shown in Figure 14.1.

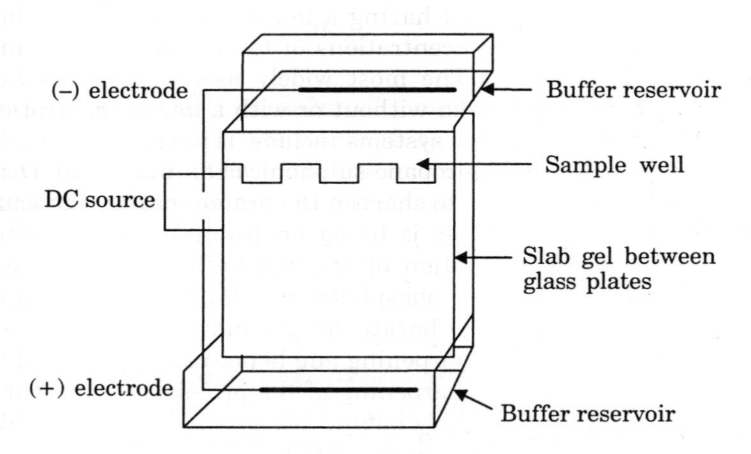

Figure 14.1 Vertical gel electrophoresis setup.

Sodium dodecyl sulphate-polyacrylamide gel electrophoresis (SDS-PAGE). This method involves electrophoretic separation of denatured proteins on polyacrylamide gel. The proteins in solution are completely denatured by treatment with the detergent sodium dodecyl sulphate (SDS) and β-mercaptoethanol and boiling the mixture for a few minutes. Addition of β-mercaptoethanol (1% v/v) disrupts disulphide linkages in the protein. SDS binds strongly to the proteins and even oligomeric proteins with polypeptide chains not bound covalently are dispersed as individual subunits. A small amount of SDS (0.1%) is sufficient to saturate the polypeptide chains. Each dodecyl sulphate group carries a negative charge and since approximately 1 SDS molecule is bound for 2 amino acid residues, the polypeptide chain gets a large negative charge. Consequently, the charge/size ratio is almost the same for all polypeptide chains and separation of the polypeptides can occur only due to the molecular sieving through the pores of the gel. The electrophoresis is carried out in vertical slab polyacrylamide gel using buffers such as tris HCl-glycine and tris-tricine buffer for small molecular weight proteins. The experimental set-up is similar to that used for native gel electrophoresis.

The method has a high resolution and gives sharp zones. The method has two distinct advantages over native gel electrophoresis. Aggregates and insoluble particles, which tend to block the pores in native gel electrophoresis are completely solubilized and converted to single polypeptides. A second advantage is that the electrophoretic mobility is related to the polypeptide size and hence it is possible to determine the molecular weight of each component.

The molecular weight of sample polypeptide chains can be determined by comparing their mobility with standard polypeptides whose molecular weight is known. A linear relationship occurs between the mobility and the log of molecular weight as given by

$$v' = v_0 \frac{A - \log M}{A} \tag{14.5}$$

where A is the log molecular weight of a molecule that has no mobility in the gel and determined by extrapolation of the straight line over the whole range and v_0 is the mobility of a small molecule unaffected by the gel and is equal to $z/6\pi\eta r$. Substituting for v_0 and rewriting, the above equation becomes:

$$v' = \left(\frac{z}{6\pi\eta r}\right) \frac{A - \log M}{A} \tag{14.6}$$

The effective viscosity of gel medium is $A\eta/(A - \log M)$.

Molecular weights of proteins may also be determined by using Ferguson plot, which is a straight line obtained by plotting the relative migration values R_f of proteins versus log molecular weight of non-denatured globular proteins. The R_f values are determined by measuring the migration distances of the proteins and dividing by the migration distance of the marker stain (bromophenol blue).

Urea gel electrophoresis. This method is used particularly for proteins that are insoluble at low ionic strength. The proteins are solubilized by denaturing them completely with urea in the presence of mercaptoethanol to disrupt any disulphide linkages. On electrophoresis in a set-up similar to that used for native gel electrophoresis, each protein is separated according to charge and subunit size. Urea containing starch gels are easier to handle compared to polyacrylamide-urea gels.

14.2.4 Practice of Preparative Gel Electrophoresis

Horizontal and vertical slab preparative gel electrophoretic methods are commonly used.

A schematic diagram of a horizontal electrophoretic setup is shown in Figure 14.2.

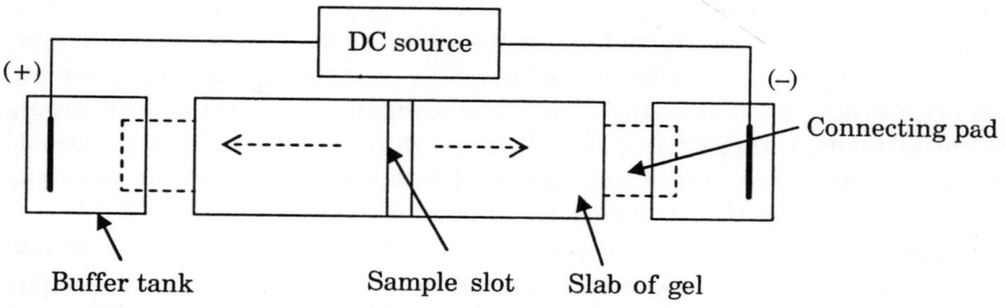

Figure 14.2 Horizontal preparative electrophoresis unit.

The apparatus consists of the separating medium (buffer mixed with gel) connected to two electrode tanks by means of gauze pads or wicks. The tanks consist of two compartments, an electrode compartment containing platinum electrode and the buffer compartment. A separate electrode compartment is necessary to minimize changes in pH close to the support medium as such changes occur in the region of the electrodes even in a buffered solution. The horizontal slab is cast from a thick slurry of starch powder, Sephadex G-25, or agarose in buffer solution and draining off the excess buffer. The ends of the slab are supported by wet cloth or paper for making contact with the buffer in the electrode tanks. The sample protein mixture in the same buffer is poured into the vertical slot cut in the slab and electrophoresis is carried out by switching on the current. The entire unit may be placed in a cold room and the current density kept low to avoid convective heating. After electrophoresis, the slab is cut into slices and the separated protein bands collected by washing the slice with buffer.

Vertical gel electrophoresis for preparative scale uses a gel column to perform the electrophoresis and the proteins migrating out of the gel are collected in a perpendicular flow of an eluting buffer.

14.2.5 Immunoelectrophoresis

Immunoelectrophoresis finds use in the identification and quantification of the complex mixture of antigens in body fluids by first separating them by electrophoresis in agar gels followed by immunoprecipitin reaction on the same gel. Immunoprecipitin reaction is a specific reaction between an antigen and its corresponding antibody. An antigen is a foreign substance, which on introduction into an animal body induces the formation of antibody. The antibody is a plasma protein belonging to the family of immunoglobulins. The addition of an antigen to a constant amount of antibody produces the antigen-antibody complex and appears as an opaque precipitate in the pH range of 7–9. The size of the precipitate grows with increasing amounts of antigen (called the antibody excess zone) added till the equivalence point. Further increase in the amount of the antigen does not increase the size of the precipitate in the equivalence zone while excess of antigen causes the precipitate to dissolve (antigen excess zone). The antigen of the sample on being subjected to electrophoretic migration in a gel containing specific antibodies to the antigen, forms antigen-antibody complexes which migrate further till they reach the equivalence point. At the equivalence point precipitation occurs and further migration does not occur. This results in the formation of rocket shaped precipitate (hence the name rocket immunoelectrophoresis) whose area is proportional to the concentration of the antigen in the sample.

14.3 CAPILLARY ELECTROPHORESIS

Capillary electrophoresis (CE) is also known as capillary zone electrophoresis (CZE) or high performance capillary electrophoresis (HPCE). It is an analytical technique requiring only micro to nano scale samples. In this technique, the components of the sample mixture are transported through a buffer filled capillary tube by a high dc potential applied across the length of the capillary tube. A schematic diagram of the experimental setup is shown in Figure. 14.3.

A fused silica capillary tube of about 50–100 cm length and an inner diameter of 25–100 μm is filled with the chosen buffer and placed between buffer reservoirs which also contain the platinum foil electrodes connected to high voltage dc source. The electrode connected to the high potential side is usually placed inside a protective plexiglass box as a safety measure against electric shock. A few nanolitres of the sample mixture is injected into the positive end of the capillary tube. A high voltage of about 20–30 kV is applied across the capillary tube through platinum foil

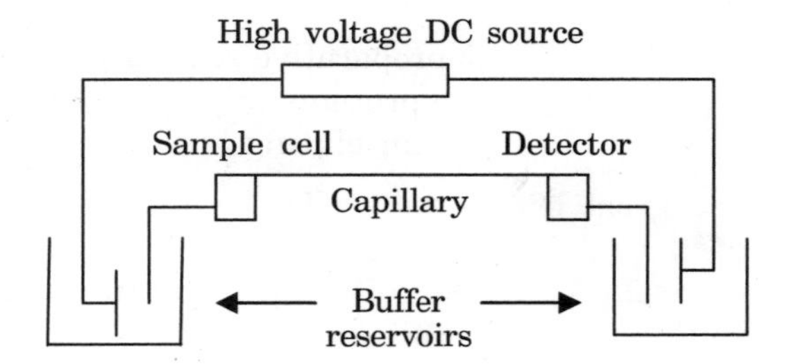

Figure 14.3 Capillary electrophoresis setup.

electrodes. The components of the mixture migrate under the influence of the applied electric field towards the negative electrode passing through a detector (e.g. UV spectral detector or fluorescence detector) placed between the end of the capillary and the buffer reservoir. The sample components get separated depending on their electrophoretic mobilities and the electroosmotic flow rate of the solvent. Electroosmotic flow of the solvent occurs in which the solvent moves from the buffer reservoir containing the positive electrode to the one containing the negative electrode due to the formation of an electrical double layer at the silica-solution interface. The fused silica surface develops fixed negative charges due to the dissociation of functional groups. The fixed negative charges attract positive ions from the buffer giving rise to an electrical double layer. The mobile positive ions are attracted towards the negative electrode and carry solvent molecules with them resulting in the electroosmotic flow of the solvent. Separation of charged species occurs as positively charged species move through the capillary at a rate greater than the electroosmotic flow rate while negatively charged species move slower than the electroosmotic flow rate or may even flow in the opposite direction as they are repulsed by the negative electrode. Neutral species move through the capillary at the electroosmotic flow rate. The high performance characteristic of the technique is due to the high resolution with the sample peaks appearing as narrow and sharp bands.

Band broadening is almost negligible with the widths of the sample peaks approaching the theoretical limits set by longitudinal diffusion because of two factors, namely, electroosmotic flow within the capillary and minimization of any thermally driven convective mixing. The unique feature of electroosmotic flow is the nearly flat flow profile as shown in Figure 14.4 compared to the hyperbolic flow profile under hydrodynamic

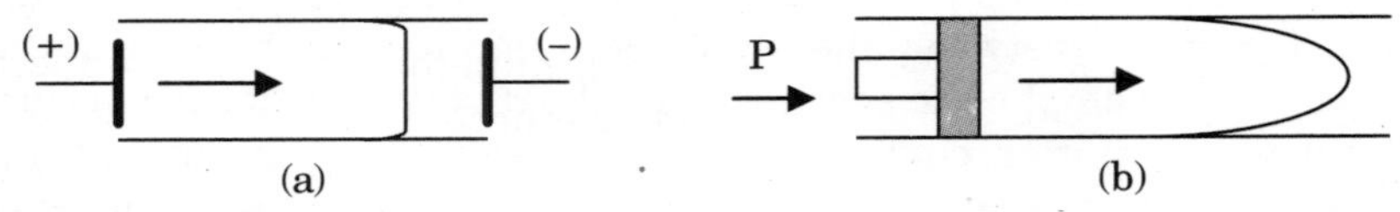

Figure 14.4 (a) Electroosmotic flow and (b) hydrodynamic flow of solvent.

pressure in column chromatography. As the flow profile is flat, band broadening is relatively small. Band broadening is also minimum in capillary electrophoresis due to negligible convective mixing. The electrical resistance along the capillary is very high because of the small cross-sectional area and great length and hence the power dissipation (which causes joule heating of the solution) is very much reduced. Any heat that is generated electrically is quickly dissipated due to the large surface to volume ratio of the capillary and hence thermally driven convective mixing is negligible.

A typical capillary electrophoretic separation of peptides at nanogram scale is completed within 10 minutes with an efficiency in terms of over 200,000 theoretical plates giving well resolved peaks similar to chromatographic peaks.

14.4 ISOELECTRIC FOCUSSING

Isoelectric focussing involves the separation of amphoteric substances (e.g. amino acids, peptides and proteins) by allowing the substances to migrate in an electric field in conjunction with a pH gradient between cathode and anode. The migration of the components of a mixture occurs up to a point in the system where the pH is equal to the isoelectric point of the substance. The electrodes are placed in a slab of gel or column of sucrose solution with a concentration gradient. Due to electrolysis of the aqueous medium, formation of H^+ ions at the anode and OH^- ions at the cathode occurs resulting in the anode region becoming low (acidic) pH and the cathode region high (basic) pH with a steep pH gradient formed in the bulk solution or the slab of gel between the two electrode regions. The formation and stabilization of a continuous pH gradient between the two electrodes is an important requirement of the technique. The pH gradient is maintained by the use of buffers called *carrier ampholytes*. The ampholytes are synthetic aliphatic polyamino-polycarboxyic acids available commercially in mixtures covering a wide pH range of 3–10 or various narrow bands of pH. The ampholytes have large number of both positive and negatively charged functional groups with closely spaced pK and pI values. During isoelectric focussing the amphoteric carrier ampholyte species also migrate to their respective isoelectric points. The carrier ampholytes consist of several thousands of amphoteric species having closely resembling pI values covering the entire pH range. They also have a good buffering capacity at their pI values. A single ampholyte species with an acidic pI will concentrate at its pI close to the anode in the pH gradient between the electrodes. Because of the high buffering capacity of the ampholyte species, a small pH plateau is formed around its pI. Since the buffer contains several species with varying pI values, overlapping pH plateaus will be formed at the respective pI values of the different ampholyte species and a continuous pH gradient is formed between the two electrodes as shown in Figure 14.5.

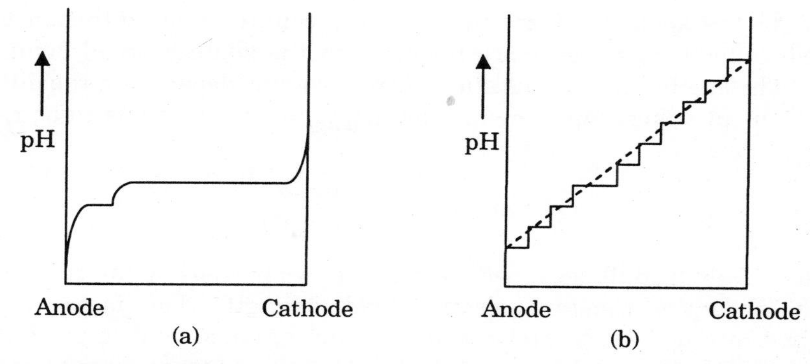

Figure 14.5 pH plateau formation by (a) a single ampholyte species and (b) overlapping pH plateaus resulting in a continuous pH gradient by several ampholyte species.

The separation of amphoteric substances by isoelectric focussing may be explained with the separation of a hypothetical sample mixture containing two proteins A and B having an acidic and basic pI respectively. The sample mixture is applied at the center of the pH gradient on the slab of gel formed by incorporating the carrier ampholyte buffer. Under the influence of the applied electric field, the proteins A and B will migrate towards the anode and cathode respectively. As the proteins approach their respective pI values their charge decreases and at pI the proteins will have no net charge and do not migrate further. Thus they get concentrated or focused at their respective pI values as shown in Figure 14.6.

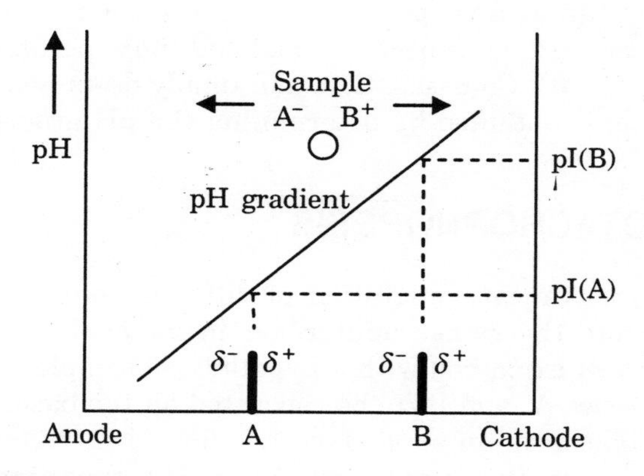

Figure 14.6 Principle of isoelectric focussing.

However, the proteins diffuse away from their pI values due to concentration differences between the regions at pI and away from pI. As they move away from the pI, they acquire charge and are forced to migrate towards the pI thereby bringing in a focussing effect. The focussing effect due to the applied electric field and the counteracting diffusion balance each other in a steady state resulting in the slightly broad band.

The resolution or separation of two proteins depends on the difference in their pI values (ΔpI) the minimum ΔpI for two proteins to be resolved being given by:

$$\Delta \text{pI} = \sqrt[-3]{\frac{D(d\text{pH}/dx)}{E(du/d\text{pH})}} \tag{14.7}$$

where D is the diffusion coefficient of the two proteins assumed to be same and E is the constant electric field strength. The factor (dpH/dx) represents a straight and continuous pH gradient and (du/dpH) is the electrophoretic mobility change with pH assumed to be same for both the proteins. Since the diffusion coefficient is inversely related to the molecular size of the proteins, large proteins tend to focus better than smaller ones. A shallow pH gradient (low value of (dpH/dx)) results in a better separation.

Analytical isoelectric focussing is usually carried out in polyacrylamide gel cast in the form of horizontal slabs. The gelling mixture contains the carrier ampholytes and an electrophoretically inert substance such as sorbitol or glycerol (10%) to increase the osmotic pressure and minimize ripples on the gel surface during the experimental run. The sample with low ionic strength is applied on the slab with the help of a plastic mask having holes. Even crude samples can be directly applied on the slab. The experiment is performed at constant power so that the establishment of the pH gradient can be followed by monitoring either the increase in voltage towards an asymptotic steady state value or the decreasing current. Once the pH gradient is established and the focussing of the sample proteins occurs at their respective isoelectric points, a steady state has been reached and the experiment concluded, the slab is stained with Comassie Blue and finally destained. The pI spectrum of the sample is obtained by determining the pH gradient formed.

14.5 ISOTACHOPHORESIS

Isotachophoresis means migration of different ionic species of the same sign all having the same counter ion in an applied electric field. The principle is best explained with a hypothetical sample mixture containing two ionic species A^- and B^- to be separated on the basis of the differences in their mobilities in an applied electric field (i.e., $m_A^- \neq m_B^-$). The sample ions have a common counter cation C^+. The experiment is carried out using (i) the sample solution and (ii) two inert electrolytes containing the ions of the same sign (anions) as the sample ions together with the common counter cation C^+. One of the electrolytes called the leading electrolyte should contain anions L^- having greater mobility than the sample ions (i.e., $m_L^- > m_A^-$ or m_B^-), while the second electrolyte called the terminating electrolyte containing anions T^- having lower mobility than the sample ions (i.e., $m_T^- < m_A^-$ or m_B^-). For example in the separation of anions in aqueous medium, the leading ion may be Cl^- and the terminating ion may be a carboxylate ion with a common counter cation of Na^+. In an applied electric field at constant current, the sample anions migrate towards the anode with a velocity, v, depending on the effective mobility of the leading anion L^- as given by the Eq. (14.8).

$$v = mE \tag{14.8}$$

where E is the electric field strength. The ions with higher effective mobility will move fast and those with lower effective mobility will follow in decreasing order of their effective mobilities. The polarity of the electric field is such that with a homogenous current density all the ions move with same speed at equilibrium and get separated into a number of consecutive zones in immediate contact with each other and arranged in order of their effective mobilities as shown in Figure 14.7. To satisfy the above condition the electric field strength E is increased when mobility, m, decreases as shown in the same figure.

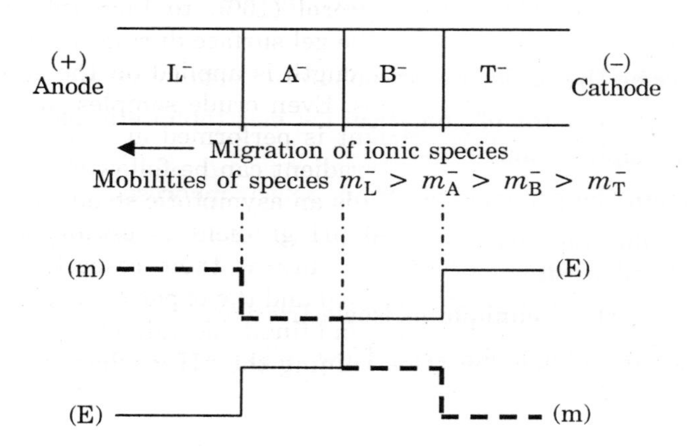

Figure 14.7 Migration of ionic species in isotachophoresis and the variation of mobilities of the species (m) and electric field strength (E) during isotachophoresis.

Kohlrausch's theoretical treatment showed that the concentrations of the ions at the migratory boundary between two electrolytes are related to their effective mobilities. The ionic concentration in a separated zone adapts itself to the concentration of the preceding zone as described by Kohlrausch's regulatory function. The ratio of the concentration C_A and C_B of the two sample ions A^- and B^- is given by the relationship:

$$\frac{C_A}{C_B} = \frac{m_A}{m_A + m_C} \frac{(m_B + m_C)}{m_B} \tag{14.9}$$

The ionic mobility is expressed in cm^2/volt sec. Since the effective mobility of each ion is constant under defined conditions we can rewrite Eq. (14.9) as

$$C_A = C_L K \tag{14.10}$$

where K is a constant. Thus, at equilibrium the sample ion concentration is directly proportional to the concentration of the leading ion. The ion concentration in each zone is therefore constant. The unique feature of isotachophoresis is that the length of the zone is directly proportional to the amount of ions in the zone which is useful in quantitative estimation of ions.

Questions

1. Give an outline of the different electrokinetic methods of separation.
2. Explain the term 'electrophoretic mobility'.
3. Discuss the theoretical principles involved in electrophoretic separation.
4. Describe the various electrophoretic techniques.
5. Give an account of analytical gel electrophoretic techniques and their applications.
6. Explain the principle of immunoelectrophoresis.
7. Describe the experimental setup and procedure of capillary electrophoresis.
8. Discuss the technique of isoelectric focussing.
9. What is isotachophoresis? How is it carried out?

CHAPTER

15 Finishing Operations and Formulation

15.1 FINISHING OPERATIONS

Biotechnological products require varying degrees of purification depending on different end uses. The different downstream processing operations that can be adopted for the isolation, concentration and purification of these products have been discussed in the earlier chapters. Many bulk products such as industrial enzymes and organic solvents are sufficiently pure at the end of these operations and ready for marketing or further conversion into end use products. However, for pharmaceuticals and many organic fine chemicals, a final finishing or polishing step is required to achieve final purification and to make the product consumer acceptable. The finishing step may involve (i) crystallization to remove closely related impurities from the desired product (ii) drying to remove any solvent including water that may be present or (iii) formulation to meet the requirements of the consumer and ensure product stability.

15.2 CRYSTALLIZATION

Crystallization is a process where solid particles of specified size and shape are formed from a homogeneous phase. It may also be considered as a solid-liquid separation process in which mass transfer of a solute occurs from the liquid to the solid phase. It is the most common method of final purification of a desired product because crystals obtained are usually of exceptional purity and free from even closely related impurities.

Crystallization yields uniform sized and shaped crystals facilitating their separation by filtration or centrifugation and drying. A crystalline product has always a better appearance and consumer acceptance.

Crystallization may be initiated either by cooling or by evaporation. The solution is concentrated either by evaporation or cooling until a saturated and finally a supersaturated solution is obtained. Saturation is the maximum concentration of the solute that is thermodynamically stable in solution. Saturation is described in the form of solubility-temperature phase diagram for individual solutes in the chosen solvent. Supersaturation results when the solution contains more solute than that is present at saturation. Supersaturated solutions are thermodynamically unstable. The degree of supersaturation of a solution is measured in terms of the supersaturation coefficient S, given by the ratio of the concentration of solute (C_t) in a solvent at a given temperature to that of the concentration of solute in solvent in a saturated solution (C_0) at the same temperature.

$$S = \frac{C_t}{C_0} \tag{15.1}$$

In a solution, the solubility is defined by the condition $S = 1$; and if $S > 1$ the solution is said to be supersaturated.

15.2.1 Crystallization Theory

The process of crystallization is considered to consist of the basic steps of nucleation and crystal growth. In the absence of any solid particle, nucleation must occur before crystal growth. The driving force for both these steps is supersaturation. The supersaturated state may be considered to consist of three loosely defined zones, a metastable zone, an intermediate zone and a labile zone. In the labile zone, nuclei are formed spontaneously from a clear solution. In the intermediate zone of supersaturation, formation of new nuclei and crystals as well as the growth of crystals occur. In the metastable zone, solute in excess of the equilibrium concentration deposits on existing crystals but no new crystals or nuclei are formed. Phase equilibrium as well as operating conditions such as speed of agitation or mixing control the supersaturated state and the three of zones of supersaturation.

Crystallization occurs only in a supersaturated solution and may be initiated by primary nucleation also called homogeneous nucleation due to rapid local fluctuations on a molecular scale in the homogeneous phase. Solute molecules come together to form clusters. Alternatively, crystallization may be initiated by heterogeneous nucleation involving the addition of solute crystals (seed) to form additional crystal nuclei (secondary nucleation). Dust, gas bubbles, mechanical shock or ultrasonic shock may also bring about heterogeneous nucleation

Crystal growth will occur subsequent to nucleation or the addition of seed material. The solubility of small crystals is greater than that of a large crystal. In a supersaturated solution, the small crystals will dissolve and large crystals will grow. The rate of growth of a crystal depends both on the transport of material to the surface of the crystal and on the mechanism of surface deposition. Stirring the solution during crystallization helps in the transport of material to the surface. However, the rate of growth of crystals is diffusion controlled. The presence of impurities usually reduces the rate of crystal growth.

The rate of growth of a crystal face is defined as the distance moved per unit time in a direction perpendicular to the face. The growth of the crystal takes place only at the outer face in a layer-by-layer process, which requires the transport of the solute from the bulk of the solution to the outer face. The solute molecules diffuse through the solution to reach the face and are integrated into the space lattice at the crystal surface (surface reaction). The diffusion and interfacial (surface reaction) steps determine the overall rate of crystal growth as given by the equation

$$\frac{dM}{dt} = kA(c - c^*) \tag{15.2}$$

where M is the crystal mass, A is the surface area of the crystal face and c and c^* are the solute concentrations in the supersaturated bulk solution and at saturation respectively. The rate constant k is the overall transfer coefficient involving both the mass transfer coefficient and the coefficient for surface reaction at the interface.

The rate equation for crystal growth involves the two parameters, namely, M and A which vary continuously and hence difficult to solve. According to McCabe all crystals that are geometrically similar and of the same material in the same solution grow at the same rate. Under such circumstances the parameters M and A may be ignored and the growth is measured on the basis of increase in length ΔL, in mm, in linear dimension of one crystal. This increase in length is for geometrically corresponding distances on all crystals and is independent of the initial size of crystals growing under the same environmental conditions. The overall transfer coefficient is the same for each face of all crystals and the growth rate G expressed as $(\Delta L/\Delta t)$ in mm/h is a constant. The ΔL law has been found to be applicable for crystal growth of many materials at crystal sizes less than 0.3 mm (50 mesh) but not applicable to systems when crystals are subjected to different treatment procedures based on their size.

15.2.2 Crystallization Practice

In industrial crystallization processes, either of the two following approaches are practiced. Finely divided solute crystals are added as seed to a solution maintained in the metastable zone of supersaturation and the seed crystals grow to finished size without further nucleation. Alternatively, nuclei may be formed either spontaneously or by secondary nucleation from the solution in the labile zone. Agitation of the solution allows the formation of new crystals in the upper range of the metastable zone.

In commercial crystallization, the yield and purity of crystals as well as the sizes and shapes of the crystals are important. Size uniformity of crystals is desirable to minimize caking in the package, for ease of pouring, for ease of washing and filtering. Sometimes, crystals of a certain shape are required such as needles rather than cubes. There are seven classes of crystals, depending on the arrangement of the constituent atoms or ions. These include the cubic, tetragonal, orthorhombic, hexagonal, monoclinic, triclinic and trigonal systems.

The external form or habit of a crystal depends on the conditions of growth. The different faces of a growing crystal grow at different rates resulting in the overlapping of the faster growing face by a slower growing face, the faster growing face ultimately being absent in the finished crystal. Crystals grown rapidly from highly supersaturated solution grow into long needles and have a dendrite (tree-like) structure because such a structure has high specific surface area and can easily dissipate the heat released during crystallization. The shape of the crystals may be selectively modified by chemicals called habit modifiers. Habit modifiers may be naturally present or deliberately added during crystallization. For example, raffinose occurs naturally in sugar beet and at concentrations of about 1% modifies the sucrose crystals to a cubic appearance, while at concentrations of about 2% modifies sucrose crystals into thin, narrow plates. Addition of a trace of potassium ferrocyanide to brine modifies sodium chloride crystals to dendrite type, an extreme form of habit modification is crystal inhibition where growth is reduced to negligible proportions on all faces. Habit modification also occurs during crystal growth when crystals come together and adhere to give twinned crystals. Crystal aggregation may also occur giving rise to large irregular shaped crystals.

The crystals formed are separated and dried to yield the finished product.

15.2.3 Equipment for Crystallization

Crystallizers may be batch or continuous type, the latter being generally preferred. Crystallization cannot occur without supersaturation. A main function of any crystallizer is to cause a supersaturated solution to form. Crystallizers bring about supersaturation either by (i) cooling the solution with negligible evaporation (ii) evaporation of the solvent with little or no cooling as in evaporator crystallizer or (iii) combined cooling and evaporation as in adiabatic or vacuum crystallizers.

Crystallizers producing supersaturation by cooling are used for solutes having solubility curves that decrease markedly with temperature.

However, in the case of common salt, the solubility curve changes little with temperature and hence evaporation of the solvent water is necessary for producing supersaturation. In adiabatic crystallizers, a hot solution is introduced into a vacuum, where the solvent evaporates and the solution is cooled adiabatically. This method is widely practiced in large scale operations. Particle size distribution of crystals is another important factor in the design of crystallization equipment.

Commercial crystallizers use different methods for contacting the supersaturated solution with growing crystals.

In the circulating magma method, the entire magma of crystals (suspension of crystals) and supersaturated liquid are circulated through the supersaturation and crystallization steps without separating the liquid from the solid. Crystallization and supersaturation occur together. The circulating magma vacuum crystallizer uses this method in which the magma is circulated out of the main body through a circulating pipe by a screw pump. The magma flows through a heater, where its temperature is raised 2–6 K. The heated liquid then mixes with the body slurry and boiling occurs at the liquid surface. This causes supersaturation in the swirling liquid near the surface, which in turn causes deposition of solid on the swirling suspended crystals until they leave again via the circulating pipe.

In a second method of contacting the supersaturated solution with growing crystals, called the circulating liquid method, a separate stream of supersaturated liquid is passed through a fluidized bed of crystals, where the crystals grow and new ones form by nucleation. Then the saturated liquid is passed through an evaporating or cooling region to produce supersaturation again for recycling. The circulating liquid evaporator crystallizer uses this method. This equipment is also called the Oslo crystallizer.

In open tank crystallizers used mostly for batch crystallization of antibiotics and other products of low molecular weight, hot saturated solutions are allowed to cool to produce supersaturation. Seed material is added to initiate crystallization. After a period of time, the mother liquor is drained and the crystals are removed. Batch crystallization yields crystals of poor quality often containing occluded mother liquor of non-uniform size.

15.3 DRYING

Drying removes the solvent, mainly water, from the desired product and in most cases stabilizes the product making it amenable for storing, packaging or formulation. Expensive organic solvents are also recovered during this step.

15.3.1 Theoretical Considerations

Non-hygroscopic materials can be dried to zero moisture level whereas most of the biological products which are hygroscopic will have a residual moisture content depending on the relative humidity of the surrounding atmosphere. The moisture content of a material as a function of relative humidity (%RH) is shown in Figure 15.1. In a hygroscopic material, the moisture may be bound within the capillaries or remain unbound within the material due to surface tension of water. Depending on the moisture content of the material and the relative humidity of the surrounding air, the material may either absorb or desorb moisture. The equilibrium moisture content is the minimum moisture to which a hygroscopic material can be dried.

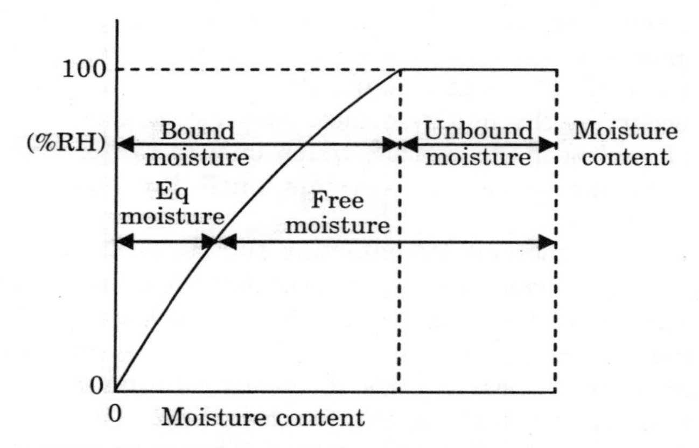

Figure 15.1 Moisture content of a material and the relative humidity of the surrounding air during drying.

The basic mechanisms involved in the rates of drying of materials are yet to be understood clearly. Hence it is necessary to obtain some experimental data on drying rates. Usually batch experiments involve the continuous monitoring of the moisture content of the sample filled in a tray so that the top surface alone is exposed to a stream of drying air. The variation of the moisture content during drying is shown in Figure 15.2.

The drying cycle consists of a number of stages as shown by the variation of drying rate as a function of free moisture as in Figure 15.3, which is more informative. The stage A-B represents the settling down period in which the solid surface comes to equilibrium with the drying air. The second stage B-C is known as constant drying rate period. During this stage the surface of solid remains saturated with liquid water because of the movement of water within solid to the surface takes place at a rate greater than the rate of evaporation from the surface. Drying takes place by movement of water vapour from the saturated surface through a stagnant air film in to the main stream of drying air. The rate of drying

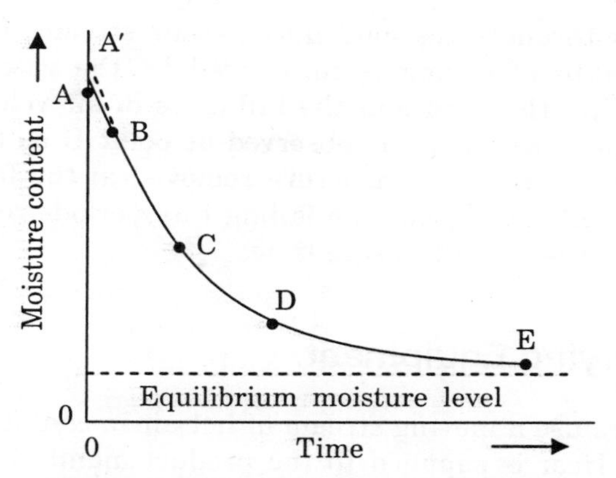

Figure 15.2 Drying rate curve.

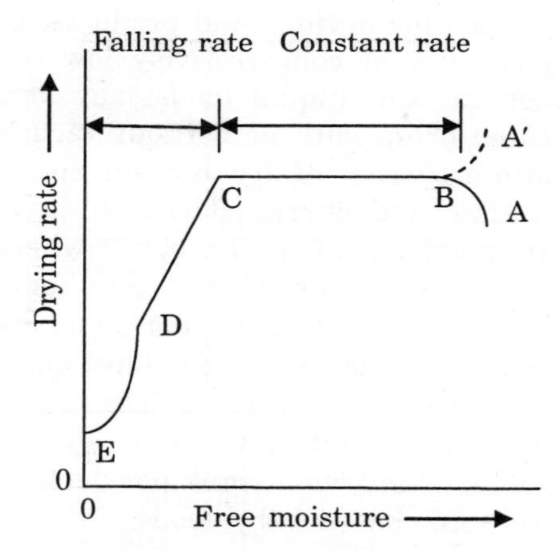

Figure 15.3 Rate of drying as a function of free moisture content.

is dependent on the rate of heat transfer to the drying surface. The rate of mass transfer balances the rate of heat transfer and hence the temperature of the drying surface remains constant. Stage C-D represents the falling rate period because as drying proceeds, the rate of movement of moisture within the material to the surface decreases and the surface begins to dry out. The moisture content of the material at point C is known as the critical moisture content.

From this point onwards, the surface temperature begins to rise as drying proceeds. Often the falling rate period consists of two parts known as the first and second falling rate periods C-D and D-E respectively. In the first falling rate period, the surface is drying out and the drying rate falls. The second falling rate period begins at point D when the surface is completely dry. The plane of evaporation slowly recedes from the surface into the solid and the drying rate falls further. Heat for the evaporation is transferred through the solid to the zone of vapourization. Vapourized water moves through the solid into the air stream. In the falling rate period, the rate of drying is influenced by the rate of movement of moisture within the solid and the influence of air velocity decreases. In some cases no sharp break is observed at point D as the surface drying is gradual. The amount of moisture removed in the falling rate periods may be relatively small, but the falling rate periods represent the major proportion of the overall drying time.

15.3.2 Drying Equipment

Hot air driers use a moving stream of hot air in contact with the product to be dried. Heat is supplied to the product mainly by convection. Kiln driers, cabinet, tray or compartment driers, tunnel driers, conveyor driers, bin driers, fluidized bed driers, pneumatic driers, rotary driers and spray driers use hot air for drying solid products. Liquid products, may be evaporated by the use of comparatively low temperatures and low pressures in a vacuum pan. Liquid foods may also be drum dried by passage over a heated drum with or without vacuum or spray dried by spraying liquids into a current of dry, hot air. Spray drier is extensively used for drying solutions and slurries. The material is introduced into the drying chamber in the form of a fine spray where it is brought into intimate contact with a stream of hot air. Very short drying time (1–10 sec) and the relatively low product temperature are the main features. The droplets of the spray usually have diameters of the order of 10–200 μm, thus presenting a large surface area per unit volume of material to the drying air resulting in rapid drying. The drier finds use in the production of skim and whole milk powders, whey solids, ice cream mix, butter, cheese, milk based baby foods, coffee, tea, dried powdered eggs, fruit and vegetable juice powders, edible proteins, yeast extracts, wheat and corn products.

Direct contact drying involves bringing the material to be dried into contact with a heated surface and heat is supplied to the product mainly by conduction. Examples include drum driers, roller driers and vacuum band driers. In such systems, the necessary sensible and latent heat of evaporation are supplied to the material by conduction. In order to achieve reasonable drying times and to dry to a low moisture content, the heated surface temperature needs to be appreciably higher than 100°C and therefore, as drying nears completion, the material temperature rises quite high. Thus the danger of heat damage to sensitive food materials is comparatively high. To reduce this hazard, drying by contact is often carried out under reduced pressure so that lower surface as well as material temperature may be employed.

15.3.3 Freeze Drying (Sublimation Drying or Lyophilization)

Foodstuffs, pharmaceuticals and biological materials which are heat sensitive even to low or moderate temperatures are freeze-dried. Freeze drying produces the highest quality product, particularly food products, with no loss of flavour and aroma and minimizes thermally induced degradation reactions. The method involves freezing of the material by exposure to cold air followed by sublimation of ice in vacuum from the frozen state to produce a dried product. Freezing leaves a mass having high structural rigidity preventing its collapse during sublimation of the ice and when water is added later, the rehydrated product retains almost the original structural form. However, the method is costly due to the low rate of drying and the necessity of high vacuum equipment to maintain low pressures.

In the freeze drying process, the material is composed of a frozen core and as the ice sublimes the plane of sublimation recedes from the outer surface leaving a porous shell of the foodstuff. The heat for the latent heat of sublimation of about 2838 kJ/kg of ice is conducted inward through the layer of dried material. The water vapour is also transferred through the layer of dried material. Thus, heat and mass transfer occur simultaneously during freeze drying.

In most of the conventional freeze-drying systems, the vapour pressure gradient necessary for sublimation is attained by maintaining the total pressure in the drying chamber of the order of 0.1–2.0 torr. A condensing system is provided to remove water vapour formed and a heating system to supply the necessary latent heat of sublimation to the frozen material. In practice it is not feasible to freeze all the liquid present in the food. The optimum rate of freezing depends largely on the material and the rate is usually determined experimentally.

Batch freeze driers. The essential components of a batch freeze drier are a vacuum cabinet, a vacuum system and a heating system. Freeze drier cabinets are essentially similar to vacuum shelf driers. The vacuum system must be capable of pumping down the cabinet pressure initially in a short time to prevent melting of the frozen products. In practice this usually entails lowering the cabinet pressure between 5–1 torr within 10 minutes. The cabinet pressure then should be brought below 1 torr and must be held at that pressure during drying. Refrigerated condensers backed by a mechanical pumping system are commonly used, commercially. Multistage steam ejectors may also be used for achieving and maintaining the required low pressure. Heat may be supplied to the frozen material by conduction or radiation or from a microwave radiator. The former two methods are used commonly, often in combination. Multi-cabinet freeze driers and tunnel driers may also be used for larger throughput.

The removal of the major portion of water by sublimation results in a product with a porous structure, retaining the shape and size of the original materials. Many of the disadvantages with other drying methods are avoided. Shrinkage is almost negligible, movement of soluble solids is limited and heat damage is minimized. The retention of volatile odour and flavour compounds in food products is also high.

Among the disadvantages is the fact that freezing damages the cell structure and the product becomes brittle. The high capital and operating costs involved in freeze-drying and stringent packaging requirements of the products make this process very expensive.

15.4 FORMULATION

Formulation, particularly of pharmaceuticals in the form of tablets, capsules, injectables, creams, powders and syrups is often required. Other biotechnological products such as bakers' yeast, industrial proteins and enzymes also require formulation depending on their end use. Formulation of a product should ensure the product quality and stability, and more importantly, the requirements of the consumer or end user being met.

Bakers' yeast obtained by aerobic fermentation is concentrated by centrifugation followed by filtration on a rotary vacuum filter to a dry matter content of about 35%. The concentrated yeast with a consistency similar to putty is extruded into long filamentous mass, broken into pieces and dried in a fluidized bed at bed temperatures between 40–60°C. Such dried yeast formulation is stable and amenable for long-term storage and transport. Soaking in water just before use restores its biological activity.

Enzymes are gaining importance as catalysts in industrial processes such as conversion of starch to glucose syrups and the production of high fructose syrup, protein hydrolysates, organic fine chemicals, food chemicals and processed food products and drug intermediates. For continuous operation of industrial processes, the enzymes need to be formulated or immobilized suitably on chosen support matrices for use in the packed bed or fluidized bed reactors. A variety of immobilization techniques are known: adsorption, gel entrapment, micro-encapsulation, ion exchange and covalent binding. It is necessary to identify surface functional groups other than those that involve in catalytic activity for immobilizing on support matrix and to select a suitable technique of immobilization depending on the requirement of the end user. For example, glucose isomerase used in the conversion of glucose to high fructose containing syrup is immobilized by cross-linking with gelatin. Penicillin acylase used in the production of semisynthetic penicillins is used in the form of immobilized dead cells. Lactase entrapped in cellulose acetate fibers is used in lactose hydrolysis in milk.

Proteases, mainly serine proteases from *Bacillus* sp. find extensive use in detergent industry, the detergent powders containing about 0.3–0.6% of proteases. The formulation process of detergents has to ensure the stability and activity of enzymes in the presence of other aggressive components of the detergents as well as operating conditions of high alkalinity and temperatures.

Questions

1. Discuss the theory of crystallization.
2. What are the advantages of crystallization as a finishing operation in bioseparations?
3. What are the theoretical considerations involved in drying of products?
4. Write a note on freeze drying and its advantages.
5. Why is formulation a necessary step in finishing operation?

Bibliography

1. Scopes, R.K., *Protein Purification—Principles and Practice*, 3rd ed., Narosa Publishing House, New Delhi, 1994.
2. Krijgsman, J., *Product Recovery in Bioprocess Technology*, BIOTOL series, Butterworth-Heinemann Ltd., Oxford, 1992.
3. Belter, P.A., E.L. Cussler and Wei-Shou Hu, *Bioseparations—Downstream Processing for Biotechnology*, John Wiley & Sons, New York, 1988.
4. Jan-Christer Janson and L. Ryden, (Eds.) *Protein Purification–Principle, High Resolution Methods and Applications*, VCH Pub., 1989.
5. Weatherley, L.R., (Ed.) *Engineering Processes for Bioseparations*, Butterworth-Heinemann Ltd., Oxford, 1994.
6. Asenjo, J.A., (Ed.) *Separation Processes in Biotechnology*, Marcel Dekker Inc., New York, 1990.
7. Geankoplis, C.J., *Transport Processes and Unit Operations*, 3rd ed., Prentice-Hall of India, New Delhi, 2002.
8. Bailey, J.E., and D.F. Ollis, *Biochemical Engineering Fundamentals*, 2nd ed., McGraw Hill, New York, 1986.
9. *Reversed Phase Chromatography—Principles and Methods*, Amersham Pharmacia Biotech.
10. *Ion-Exchange Chromatography—Principles and Methods*, Amersham Pharmacia Biotech.
11. *Hydrophobic Interaction Chromatography—Principles and Methods*, Pharmacia.
12. *Isoelectric Focusing—Principles and Methods*, Pharmacia Fine Chemicals.
13. *Gelfiltration—Theory and Practice*, Pharmacia Biotech.
14. *Isotachophoresis*, LKB Produkter AB.

Index

Adsorption, 54–64
  batch mode, 56–57
  CSTR mode 57–60
  fixed bed 60–64
  isotherms 54-56
Affinity chromatography, 214–215
  advantages, 215
Affinity partitioning, 89
Aqueous two-phase extraction, 81–89
  applications, 88–89
  equipment, 87
  phase diagram, 82–84
  principles, 85–86
  process, 86–87

Basket centrifuge, 50–52
Batch
  adsorption, 56–57
  extraction, 76–78
Bead mill, 20–22
Bingham model, 7, 8
Bioaffinity chromatography, 215–227
  practice, 218–222
  principles, 215–218
  spacer arm, 219–220
Bioprocess products, 2, 3

Capacity factor, 152
Capillary electrophoresis, 248–250
Capillary flow model, 102
Cell wall disruption methods, 13–25
  bead mill, 20–22
  chemical methods, 16–20
  detergent solubilization, 17
  homogenization, 22–23
  mechanical methods, 20–24
  permeabilization, 17, 18
  osmotic shock, 15
  sequential disruption, 19, 20
  thermolysis, 15
Centrifugal force, 42
  filtration, 50–52
  decantation, 45
  sedimentation, 43–44
Centrifugation, 42–52
Centrifuges, 45–47
  disc (stack) bowl, 47
  multi-chamber, 47
  scale-up, 50
  selection, 49
  tubular bowl, 45–47
Chemical potential, 69, 70, 128, 129
Chemistry of thiol groups, 233
Chromatofocusing, 196–199
Chromatography
  classification, 143–145
  scale up, 166–167
Chromatotopography, 207
Cohn equation, 121, 123
Co-current extraction, 79
Column
  chromatography description, 145–146
  efficiency, 153
Concentration polarization, 104–105
Continuous extraction, 79–81
Counter-current extraction, 79

Covalent chromatography, 214, 232–238
principles, 234–235
practice, 235–237
Cross flow filtration, 110–112
Crystallization, 255–259
equipment, 258–259
practice, 257–258
theory, 256
CSTR adsorption 57–60

Decaffeination of coffee, 96
Dialysis, 101, 102, 104, 116–117
Displacement analysis, 148
Dissociative extraction, 71–72
Drying, 259–262
equipment, 262
theory, 260–261
Dye-protein interaction, 228–229

Eilers' equation, 7
Einstein equation, 7
Electrodialysis, 102, 117
Electrophoresis, 240–248
concept, 241–242
techniques, 242–248
Elution analysis, 146
Extraction, 67–96
equipment, 74–75
operating modes, 76–80
principles, 68–70
process, 71
selectivity, 70

Fermentation broth characteristics, 4–8
Fermentation broth pretreatment, 30–32
Filter
aids, 31
media, 32
press, 33
Filtration, 26–37
equipment, 32–33
theory, 27–28
Fixed bed adsorption, 60–64
breakthrough curve, 60
scale up, 62–64
Flocculation, 30, 140
Formulation, 255, 264–265
Fouling of membranes, 105
Freeze drying, 263–264
Frontal analysis, 148

Gaussian peak, 153
Gel filtration, 175–186
applications, 182–186
desalting by, 185
estimation of molecular weight by, 183–184
materials, 180–181
practice, 181–182
principles, 175–179
Gradient elution, 162, 163

HEEP, 155
Hetaeric chromatography, 206
Hofmeister series, 123, 201
Hollow fibre membrane module, 109
Homogeneizer, 22, 23
HPLC, 163–166
HPTLC, 169
Hydrophobic interaction, 200–201
chromatography 200, 209–213
Hyperfiltration, 114

Immobilization of ligands, 224–227
Ion exchange chromatography, 188–196
applications, 195–196
operating modes, 192–194
practice, 194–195
principles, 188–189
Ion exchangers, 189–190
capacity, 190–191
Ion-pair extraction, 73
Isocratic elution, 162
Isoelectric focusing, 250–252
Isoelectric precipitation of proteins, 121
Isotachophoresis, 252–254

Langmuir adsorption, 55
Liquid–liquid extraction, 67
Lyophilization, 263–264
Lyotropic series, 123, 201

Membrane modules, 108–110
flat sheet, 109, 110
hollow fibre, 109, 110
spiral wound, 109, 110
tubular 109, 110
Membrance separation processes, 100–118
classification, 101
factors affecting separation, 103–106
merits, 100
Microfiltration, 110–112

Newtonial fluids, 6, 7, 8
Non-Newtonian fluids, 6, 7, 8
Number of theoretical plates, 154–156

Orthokinetic aggregation, 138, 139–140
Osmotic effect, 114–115
Osmotic shock, 15

Paper chromatography, 167, 168
Partition coefficient, 68, 149
Partitioning of cell debris, 89
Peak asymmetry, 157
Peak broadening, 160
Perikinetic growth, 138
Pervaporation, 117
Physical extraction, 71
Protein precipitation, 119–141
classification of methods, 120
by metal ions, 133
by organic solvent, 127–130
by non-ionic polymers, 131
by polyelectrolytes, 132
by salts, 122–127
by selective denaturation, 133–137
in large scale, 137–141

Reactive extraction, 73
Rejection coefficient, 103
Relative retention, 152
Resolution, 155–157
Retention time, 149–151
Retention volume, 151
Reverse osmosis, 101, 102, 104, 114–116
Reversed micellar extraction, 90
Reversed phase chromatography, 200, 202, 203–209
applications, 207–209
basic theory, 202–203
practice, 202–207

Salting-in of proteins, 122
Salting-out of proteins, 122–125
SDS-PAGE, 246
Selective
denaturation of proteins, 133–137
extraction, 72–73
Smoluchowski's theory, 138
Supercritical fluid extraction, 91–96
applications, 95–96
principles, 93–94
process, 95
solvents, 91

Theoretical plates, 154–156
Thermolysis, 15
Thin layer chromatography, 167, 169
Tortuosity factor, 159
Tubular bowl centrifuge, 45–47

Ultrafiltration, 101, 102, 104, 112–114
Ultrasonication, 16

van Deemter equation, 159
plot, 160
van't Hoff equation, 15
van der Waals' interactions, 202